FORESTRY COMMISSION BULLETIN 98

Monitoring of Forest Condition in Great Britain – 1990

J. L. Innes and R. C. Boswell
Forestry Commission

LONDON: HMSO

© *Crown copyright 1991*
First published 1991

ISBN 0 11 710298 9
ODC 425.1 : 423.1 : 181.45 : 524.6 : (410)

KEYWORDS: Tree health, Pollution, Forestry

Enquiries relating to this publication
should be addressed to:
The Technical Publications Officer,
Forestry Commission, Forest Research Station,
Alice Holt Lodge, Wrecclesham,
Farnham, Surrey, GU10 4LH

Front cover: Composite silhouette of two Norway spruce (*Picea abies*) of different crown habits, showing the difference in light transmission. Left half, brush type; right half, comb type.

Contents

	Page
Summary	v
Résumé	vii
Zusammenfassung	ix
Introduction	1
Distribution of plots	1
Assessment procedures	3
Quality assurance	5
Changes during the assessment period	10
Results	17
Recognition of anomalous or unusual records	41
Environmental factors affecting crown condition	43
Pests and pathogens in 1989–90	44
Discussion	44
Conclusions	50
Acknowledgements	51
References	51

Monitoring of Forest Condition in Great Britain – 1990

Summary

A total of 7644 trees were assessed in the main Forestry Commission monitoring programme in 1990. Five species were examined: Sitka spruce (*Picea sitchensis*), Norway spruce (*P. abies*), Scots pine (*Pinus sylvestris*), oak (*Quercus* spp.) and beech (*Fagus sylvatica*), distributed over 319 sites. This represents an increase on 1989, and was attributable to an increase in the number of oak plots assessed.

Data completeness was high, exceeding 97% in each species. There was a lack of consistency in the way in which individual observers assessed some of the indices. This limited the manner in which the data could be analysed. For example, differences between observers meant that it was not possible to produce reliable maps of specific indices of forest condition. However, assessments of one of the most important of the indices, crown density, were considerably more consistent between surveyors than in previous years.

Overall, the crown density of Sitka spruce and oak improved, Norway spruce and Scots pine remained the same and beech deteriorated. However, these broad trends conceal a dynamic situation, with individual trees varying in their changes between 1989 and 1990. The improvement of crown density in Sitka spruce was due to the recovery of trees following severe defoliation by the green spruce aphid in 1989. The high level of insect activity can be attributed to the unusually mild winter of 1988–89 and although the winter of 1989–90 was similar, insect activity was not as great.

An increase in the amount of dieback in both oak and beech was recorded in 1990. In beech, dieback was accompanied by a number of other symptoms of stress, including leaf-rolling, heavy fruiting, premature leaf loss and abnormally small leaves on many trees. Water stress associated with the summer droughts of 1989 and 1990 was the most likely factor causing the symptoms. Coning in the two spruce species was particularly abundant and may also be related to the drought. Very little discoloration was seen, particularly in the broadleaves, although there was a small increase in the amount of yellowing of current-year Sitka spruce needles and the amount of browning of current-year Norway spruce needles. As with crown density, analysis of the changes on individual trees indicated a complex and dynamic pattern that is not reflected in the overall changes in condition.

Over the four years for which data are available, no indication of a large-scale decline in forest condition has been identified. There is a large amount of variation from year-to-year, with individual species reacting differently to the range of factors that affect forests in any given year. This makes identification of long-term trends difficult, and several more years of information are required before any such trends can be recognised.

The storms of January and February 1990 caused some structural damage to broadleaves in the south and south-east of England. However, the effects of the damage on crown density were confounded by other, more widespread, trends.

It is now possible to identify anomalous sites and trees with some degree of confidence. A variety of indices can be assessed and these provide the basis for recognising a tree in good or poor condition. The technique is currently being developed further and will be used in the analysis of future data.

Contrôle de la Condition des Forêts en Grande-Bretagne – 1990

Résumé

Au total, 7644 arbres ont été étudiés en 1990, dans le cadre du programme principal de contrôle lancé par la Forestry Commission. Cinq espèces ont été examinées: l'épicéa Sitka (*Picea sitchensis*), l'épicéa commun (*P. abies*), le pin sylvestre (*Pinus sylvestris*), le chêne (*Quercus* spp.) et le hêtre (*Fagus sylvatica*), tous ces arbres se trouvant dans 319 stations différentes. Ce nombre est plus important qu'en 1989 et s'explique par une augmentation des terres à chênes dans le cadre de l'étude.

Le taux d'informations collectées a été très élevé, dépassant 97% pour chaque espèce. Mais un manque d'uniformité a été observé dans la manière dont les observateurs ont évalué certains critères. En conséquence, l'analyse des informations a été quelque peu restreinte. Par exemple, les différences existant entre les observateurs ont rendu impossible la réalisation de cartes fiables montrant des critères spécifiques propres aux forêts. Cependant, l'un des critères les plus importants, la densité des couronnes, a été étudié par les chercheurs de manière beaucoup plus cohérente par rapport aux années précédentes.

Globalement, la densité des couronnes de l'épicéa Sitka et du chêne s'est améliorée. Celle des épicéas communs et des pins sylvestres est restée stable et celle des hêtres s'est détériorée. Cependant, ces tendances générales cachent une situation dynamique et les changements observés parmi les arbres individuellement, diffèrent de 1989 à 1990. L'amélioration observée dans la densité des couronnes de l'épicéa Sitka s'explique par le rétablissement des arbres après la défoliation importante subie par les aphides en 1989. Le rôle très important des insectes peut être atribué à l'hiver très doux de 1988 et bien que l'hiver de 1989 ait été similaire, l'activité des insectes n'a pas été aussi importante.

En 1990, on a assisté à un dépérissement plus important des chênes et des hêtres. Pour les hêtres, ce dépérissement s'est accompagné d'un certain nombre de symptômes dûs aux contraintes exercées par l'environnement, d'où le roulement des feuilles, la production excessive de fruits, la perte prématurée des feuilles et la pousse de feuilles anormalement petites sur de nombreux arbres. Ces symptômes s'expliquent principalement par le manque d'eau lié aux sécheresses de 1989 et 1990. La formation de cônes a été très fréquente pour les deux espèces d'épicéas et a peut-être été due également à la sécheresse. Le phénomène de décoloration s'est très peu manifesté, bien que le jaunissement ait frappé un peu plus les aiguilles nouvelles de l'épicéa Sitka et le brunissement les aiguilles les plus jeunes de l'épicéa commun. Comme ce qui est le cas pour la densité des couronnes, l'analyse des changements observés sur les arbres individuels a indiqué une évolution complexe et dynamique qui n'apparaît pas dans les changements de l'état global des forêts.

Au cours des quatre années pour lesquelles des informations sont disponibles, aucun déclin majeur de l'état des forêts n'a été relevé. Un grand nombre de variations apparaissent d'année en année, certaines espèces particulières réagissant différemment face aux facteurs qui affectent les forêts de manière générale. Cela rend plus difficile l'identification de certaines tendances à long terme, et plusieurs années d'informations sont nécessaires avant de pouvoir établir ce genre de tendances.

Les tempêtes survenues en janvier et en février 1990 ont endommagé sérieusement les arbres feuillus du sud et du sud-est de l'Angleterre. Cependant, les effets des dégâts affectant la densité des couronnes ont été surpassés par d'autres tendances beaucoup plus répandues.

Il est maintenant possible d'identifier avec une quasi-certitude les stations anomaux et les arbres anomaux. Un certain nombre de critères peuvent être étudiés et ils permettent de déterminer quels sont les arbres en bonne ou mauvaise condition. Cette technique est actuellement en cours de développement et sera utilisée dans l'analyse des informations à l'avenir.

Überwachung des Forstzustandes in Großbritannien – 1990

Zusammenfassung

Ingesamt 7644 Bäume wurden 1990 im Hauptüberwachungsprogramm der Forstverwaltung beurteilt. Fünf Baumarten wurden untersucht: Sitkafichte (*Picea sitchensis*), Fichte (*P. abies*), Kiefer (*Pinus sylvestris*), Eiche (*Quercus* spp.) und Rotbuche (*Fagus sylvatica*), die über 319 Standorte verteilt waren. Das ist mehr als 1989 und der Grund dafür ist, daß mehr Eichenflächen beurteilt wurden.

Der Grad an Datenvollständigkeit war hoch und bei jeder Baumart über 97%. Ein Mangel an Einheitlichkeit wurde bei der Art, auf die individuelle Beobachter manche der Kategorien beurteilten, festgestellt. Dadurch war die Methode der Datananalyse begrenzt. Zum Beispiel bedeuteten Unterschiede zwischen Beobachtern, daß es nicht möglich war, zuverlässige Karten über bestimmte Kategorien der Forstzustände anzulegen. Die Beurteilung einer der wichtigsten Kategorien, Kronendichte, wies jedoch eine größere Einheitlichkeit zwischen den Beobachtern als in vergangenen Jahren auf.

Im großen und ganzen verbesserte sich die Kronendichte der Sitkafichte und Eiche, Fichte und Kiefer blieben gleich und Rotbuche verschlechterte sich. Diese breiten Tendenzen verbergen jedoch eine dynamische Situation, wobei sich individuelle Bäume zwischen 1989 und 1990 verschieden veränderten. Die Verbesserung der Kronendichte der Sitkafichte geschah aufgrund der Genesung von Bäumen, die 1989 wegen der Fichtennadellausplage eine starke Entlaubung erlitten hatten. Der hohe Grad an Insektenaktivität kann auf den ungewöhnlich milden Winter 1988 zurückgeführt werden, und obwohl der Winter 1989 ähnlich war, war die Insektenaktivität weniger groß.

1990 wurde ein zunehmendes Absterben von Eiche und Rotbuche festgestellt. Bei der Rotbuche wurde das Absterben von weiteren Belastungssymptomen begleitet, wie z.B. Blattrollen, Fruchtreichtum, vorzeitiger Blattverlust und abnormal kleine Blätter an manchen Bäumen. Der durch die trockenen Sommer von 1989 und 1990 entstandene Wassermangel war höchstwahrscheinlich der Grund für diese Symptome. Die Zapfenbildung bei den zwei Fichtenarten war besonders reichlich und kann ggf. auch der Dürre zugeschrieben werden. Besonders an den Laubbäumen wurde sehr wenig Verfärbung festgestellt, obwohl der Grad an gelblicher Verfärbung an den diesjährigen Nadeln der Sitkafichten und an bräunlicher Verfärbung der diesjährigen Nadeln der Fichten leicht erhöht war. Wie bei der Kronendichte wies auch die Analyse der Veränderungen an individuellen Bäumen ein kompliziertes und dynamisches Muster auf, das sich nicht in der allgemeinen Veränderung des Zustandes widerspiegelt.

In den vier Jahren, für die Daten zur Verfügung stehen, konnte kein Anzeichen einer Verschlechterung des Forstzustandes in größerem Maße festgestellt werden. Von Jahr zu Jahr gibt es große Unterschiede, und einzelne Baumarten reagieren verschieden auf diejenigen Faktoren, die Forstgebiete in individuellen Jahren beeinflussen. Dadurch wird die Identifizierung von langfristigen Trends erschwert, und es werden Informationen über viel mehr Jahre benötigt, bevor solche Trends erkannt werden können.

Die im Januar und Februar 1990 aufgetretenen Stürme verursachten einen gewissen strukturellen Schaden an Laubbäumen im Süden und Südosten von England. Die Auswirkungen der Beschädigung der Kronendichte wurden jedoch durch andere, weiter verbreitete, Trends verwirrt.

Es ist nun möglich, anomale Standorte und Bäume mit einem gewissen Grad an Zuversicht zu identifizieren. Verschiedene Kategorien können bewertet werden und diese bieten die Basis dafür, einen Baum in gutem oder schlechtem Zustand erkennen zu können. Diese Methode wird zur Zeit weiterentwickelt und bei der Analyse von zukünftigen Daten verwendet.

Monitoring of Forest Condition in Great Britain – 1990

J. L. Innes and R. C. Boswell, *Forestry Commission*

Introduction

Each year, the Forestry Commission examines a sample of trees for signs of poor condition. Two separate surveys are undertaken, one using a systematic, grid sampling design and the other being much more selective. This report is primarily concerned with the latter. Surveys of forest condition have been undertaken by the Forestry Commission since 1984. During this period, substantial changes have been made to the way in which trees are assessed as knowledge on the most appropriate methods of assessment has developed. These changes effectively mean that it is impossible to make comparisons of the condition of trees over the seven year period. Instead, 1987 has been used as a base year, as in previous reports (Innes and Boswell, 1987, 1989, 1990a).

During the period, there have also been changes in emphasis. The programme was initiated in 1984 in reponse to widespread concern that the forests of Britain could be adversely affected by 'acid rain', with the effect taking the form of a progressive decline in the condition of forests, ultimately leading to the collapse and destruction of forest ecosystems. Surveys in Britain and elsewhere have not so far revealed any untoward decline in tree condition, although the condition of forest ecosystems in some parts of Europe continues to decline.

Although it is generally felt that a widespread and catastrophic decline in health is unlikely in Britain, monitoring has continued as it has been recognised that the basic information provided by the programme will provide important information for evaluating the effects of environmental change. For example, changes in the chemical composition of the atmosphere may lead to changes in forest condition. The programme will provide essential information about changes in forest condition over longer periods and this will help to determine any effects of global environmental change. In addition, the programme is beginning to provide useful information on the responses of forests to natural stresses. During the winter of 1988–89 Sitka spruce was adversely affected by the green spruce aphid (*Elatobium abietinum* Walker; Hemiptera) in many areas and the programme is revealing the nature, rate and extent of the subsequent recovery. Similarly, severe droughts occurred in some areas in 1989 and 1990 and the programme has been able to assess the effects of these on forest trees, in terms of both crown condition and growth.

Distribution of plots

The distribution of plots of each species in 1990 is given in Figure 1. The distributions are similar to 1989, with the exception of oak, which had an increased sample size. Nine plots were lost as a result of storm damage, endemic windthrow or felling operations and, in contrast to previous years, none of these was replaced. The number of oak plots was increased from 53 to 73. In total, there were 56 Sitka spruce plots, 73 Norway spruce plots, 81 Scots pine plots, 73 oak plots and 36 beech plots.

The lack of plots in the south-east is the result of the storms of 1987 and 1990, when a number of stands used in the programme were lost. Sitka spruce is relatively rare in the south and east of the country and this is reflected in the sampling design. Similarly, beech is

Figure 1. Location of plots used in the 1990 programme.

a. Sitka spruce
b. Norway spruce
c. Scots pine
d. Oak
e. Beech

increasingly rare towards the north of the country and the sampling density indicates this trend.

The aim of the sampling programme is to achieve an even coverage of plots in the main areas of each species in Great Britain. Analyses of data from previous years (Innes and Boswell, 1987, 1989, 1990a,b) indicate that a sample size of about 80 plots is required to achieve a sampling density suitable for the types of spatial and temporal statistical analyses that are undertaken. Consequently, the sample sizes of most species will be increased as the availability of resources permit.

Table 1. Indices assessed in the 1990 programme.

Variable	Sitka spruce	Norway spruce	Scots pine	Oak	Beech
Branch density	x	x	x		
Branch pattern	x	x			
Browning of current needles	x	x	x		
Browning of leaves				x	x
Browning of older needles	x	x	x		
Butt damage	x	x	x	x	x
Canopy closure around tree	x	x	x	x	x
Crown density	x	x	x	x	x
Crown dieback extent				x	x
Crown dieback location				x	x
Crown dieback type				x	x
Crown form			x	x	x
Diameter at breast height	x	x	x	x	x
Defoliation type	x	x	x	x	x
Degree of leaf-rolling					x
Epicormics on branches				x	
Epicormics on the stem				x	
Flowering in lower crown			x		
Flowering in upper crown			x		
Frequency of leaf-rolling					x
Fruiting	x	x	x	x	x
Fungal presence	x	x	x	x	x
Insect damage	x	x	x	x	
Leader condition	x	x	x		
Leaf size					x
Mechanical damage to crown	x	x	x	x	x
Needle retention	x	x	x		
Overall discoloration	x	x	x	x	x
Premature leaf loss					x
Secondary shoot frequency	x	x			
Secondary shoot location	x	x			
Shoot death extent	x	x	x		
Shoot death location	x	x	x		
Stem damage	x	x	x	x	x
Tree dominance	x	x	x	x	x
Type of leaf browning				x	x
Type of leaf yellowing				x	x
Type of needle yellowing	x	x	x		
Yellowing of current needles	x	x	x		
Yellowing of leaves				x	x
Yellowing of older needles	x	x	x		

Assessment procedures

Detailed assessment procedures have been described by Innes and Boswell (1990a) and Innes (1990). A list of the indices assessed on each tree is provided in Table 1. The indices are assessed on 24 objectively selected trees at each site.

Trees were assessed from the ground by a single surveyor. In past years, two surveyors have been used, but this procedure was dropped in 1990. This change, which was introduced to reduce costs, resulted in an increase in the amount of variation between observers and the control team, leading to a reduction in the internal consistency of the data, but achieved its aim of reducing unit costs by about 40%.

A major change introduced to the assessment methods in 1990 was the use of hand-held microcomputers for data collection. These instruments were programmed to prompt the surveyor for each of the indices listed in Table 1. A number of hardware and software problems were encountered with their use, although these were no greater than anticipated with the use of new technology. The instruments resulted in a slightly longer time being spent in the field, but greatly reduced the number of errors associated with data preparation and the speed of data handling was improved. Further details are provided in the section on quality assurance.

Table 2. Data completeness, expressed as a percentage of the total number of records that should have been collected.

Variable	Sitka spruce	Norway spruce	Scots pine	Oak	Beech
Branch density	98	99	100		
Branch pattern	98	99			
Browning of current needles	98	99	98		
Browning of leaves				99	100
Browning of older needles	98	99	100		
Butt damage	98	99	100	98	100
Canopy closure around tree	98	99	100	98	100
Crown density	98	99	100	98	100
Crown dieback extent				98	100
Crown dieback location				98	100
Crown dieback type				98	100
Crown form			100	98	100
Diameter at breast height	98	99	100	98	100
Defoliation type	98	99	100	98	100
Degree of leaf-rolling					100
Epicormics on branches				98	
Epicormics on the stem				98	
Flowering in lower crown			100		
Flowering in upper crown			100		
Frequency of leaf-rolling					100
Fruiting	98	99	100	98	100
Fungal presence	98	99	100	98	100
Insect damage	98	99	100	98	100
Leader condition	98	98	100		
Leaf size					100
Mechanical damage to crown	98	99	100	98	100
Needle retention	98	99	100		
Overall discoloration	98	99	100	98	100
Premature leaf loss					100
Secondary shoot frequency	98	99			
Secondary shoot location	98	99			
Shoot death extent	98	99	100		
Shoot death location	98	99	100		
Stem damage	98	99	100	98	100
Tree dominance	98	99	100	98	100
Type of leaf browning				98	100
Type of leaf yellowing				98	100
Type of needle yellowing	98	99	100		
Yellowing of current needles	98	99	100		
Yellowing of leaves				98	100
Yellowing of older needles	98	99	100		

Quality assurance

As in previous years, considerable emphasis has been placed on quality assurance. This can be divided into two main aspects: data completeness and data quality.

Data completeness

Data completeness is summarised in Table 2. This table obscures the actual number of records lost from the analysis because of the large sample sizes involved. 886 out of 49 248 records for Sitka spruce were lost, and the corresponding figures for the other species were: 877 out of 63 936 for Norway spruce, 58 out of 69 984 for Scots pine, 1092 out of 55 056 for oak and 6 out of 28 512 for beech. The relatively high figures for Sitka spruce, Norway spruce and oak were the result of the loss of data from one plot each caused by a hardware problem. Although these figures seem high, they are well below the data loss level that would be unacceptable in a study of this nature (>5%).

Data quality

Data quality assurance forms an essential component of any long-term monitoring programme (Schlaepfer, 1985; Schlaepfer et al., 1985; Cline and Burkman, 1989). All surveyors received training for one week in the assessment of trees. Of the nine surveyors involved, seven had worked on the 1989 programme and one of the remaining two had been involved between 1984 and 1987. Consequently, only one person was new to the programme and he received additional training.

Training was helped by the use of a set of standard operating procedures (Innes, 1990). These have replaced the Swiss SANASILVA guide (Bosshard, 1986) used in previous years, although great care was taken to ensure standardisation between the two guides.

Field checking was undertaken by the senior author and involved visits to each of the surveyors in their respective field areas. 71 (5.3%) of the 1344 Sitka spruce trees assessed in the programme were checked and the corresponding figures for the other species were 168 (9.6%) of the 1752 Norway spruce, 198 (10.2%) of the 1944 Scots pine, 167 (9.6%) of the 1741 oak and 88 (10.2%) of the 864 beech.

The results for each species are presented in Table 3. It is difficult to examine these for statistical differences as the nature of the data is variable. For example, crown density is assessed on a linear scale from 0% to 100%, and parametric difference tests can be used. Discoloration is also assessed in percentages, but the classes used are of uneven width. Parametric tests can be used if the classes are converted to percentages, or non-parametric tests can be used on the classes. Other data, such as the location of dieback, are non-linear and non-parametric tests have to be used. In this analysis, either t-tests (for data recorded at 5% intervals or less) or Wilcoxon matched pairs tests (for categorical data) have been used. The t-tests provide an indication of any overall bias in the data but it does not reflect any bias apparent in individual observers. The Wilcoxon test indicates whether two sets of data have been drawn from different populations, but does not provide an indication of overall or individual observer bias. Identification of problems associated with particular surveyors is currently being addressed.

The majority of the data has been examined using the Wilcoxon test. This test has an important feature that should be mentioned. It is very sensitive to small variations when the majority of the observations are equal. For example, in Table 3c, 7 of the 198 trees received browning scores for older needles that were different between observers. The differences were due to the control observer scoring trees one (six cases) or two (one case) classes above the surveyor. In relation to the overall sample, the differences are negligible, but trees with the same score are excluded from the calculation of the T value, resulting in a significant difference being recorded. A limitation of the technique (and all others) is that if a particular feature was absent from the sampled trees, differences in assessments would not be recorded.

Sitka spruce

The results for Sitka spruce (Table 3a) indicate

Table 3a. Reliability of Sitka spruce data, as indicated by comparisons between the surveyors' and the check team's scores. Sample size: 71. Neither test could be used when the number of differences was less than three. Significant ($p < 0.05$) differences between the assessments are shown in bold.

Variable	Number of differences	t-test t	Wilcoxon test T	Z	p
Dominance	6		10.5	0	1.000
Canopy closure	17		51.0	1.21	0.230
Branch pattern	8		11.0	0.98	0.329
Defoliation type	10		18.5	0.92	0.361
Crown density	6	1.653			0.103
Shoot death location	16		50.0	0.93	0.354
Shoot death extent	13		4.0	2.90	**0.004**
Secondary shoots location	12		5.5	2.63	**0.008**
Secondary shoots abundance	9		18.0	0.53	0.595
Leader condition	2		-	-	-
Coning	1		-	-	-
Branch density	12		0.0	3.06	**0.002**
Needle retention	1		-	-	-
Browning of current needles	0		-	-	-
Browning of older needles	0		-	-	-
Yellowing of current needles	3		1.5	0.80	0.424
Current yellowing type	0		-	-	-
Yellowing of older needles	3		2.0	0.54	0.594
Older yellowing type	2		-	-	-
Overall discoloration	4		5.0	0	1.000
Mechanical damage type	5		0	2.02	**0.043**
Mechanical damage %	5	−1.943			0.056
Stem damage type	1		-	-	-
Stem damage %	0	-			-
Insect activity	11		7.0	2.31	**0.021**
Fungi	0		-	-	-

that the observations for most indices were satisfactory. Differences in scoring were reported for shoot death extent, secondary shoot location, branch density, type of mechanical damage and insect activity. Shoot death extent is an important variable and the scoring problem probably arose from confusion over the exclusion/inclusion of shoots within the crown. Suppressed shoots should be excluded from the score, but it is not always easy to tell which have been killed by suppression and which have died as a result of other factors. Difficulties were experienced with the scoring of secondary shoot location when this variable was introduced in 1989; these problems have not yet been resolved. Differences in scoring for insect activity were also experienced in 1989, and can be related to the identification and inclusion of current and old *Elatobium* damage.

Norway spruce

More differences were recorded in Norway spruce than in Sitka spruce. This is surprising, as the two species are assessed in exactly the same way. Difficulties were experienced with canopy closure, defoliation type, secondary shoot location, secondary shoot abundance, branch density and stem damage type and extent. Canopy closure is a straightforward measure about which there should be no problem and it will be resolved by further training. Defoliation type is more difficult. Where scores were different, it was primarily because the control observer gave a higher score than the surveyor. The differences over the scoring of secondary shoots were also noted for Sitka spruce, as were the differences for branch density. The differences over stem damage appear to be related to observation, with the

Table 3b. Reliability of Norway spruce data, as indicated by comparisons between the surveyors' and the check team's scores. Sample size: 106. Neither test could be used when the number of differences was less than three. Significant ($p < 0.05$) differences between the assessments are shown in bold.

Variable	Number of differences	t-test t	Wilcoxon test T	Z	p
Dominance	21		77.0	1.34	0.184
Canopy closure	39		101.5	4.03	**0.000**
Branch pattern	18		81.0	0.20	0.847
Defoliation type	22		43.0	2.71	**0.007**
Crown density	34	−1.60			0.114
Shoot death location	26		101.5	1.88	0.060
Shoot death extent	27		125.5	1.53	0.130
Secondary shoot location	27		41.5	3.54	**0.000**
Secondary shoot abundance	24		69.5	2.30	**0.021**
Leader condition	16		50.5	0.90	0.368
Coning	13		32.5	0.91	0.365
Branch density	25		0	4.37	**0.000**
Needle retention	17		53.5	1.09	0.279
Browning of current needles	1		-	-	-
Browning of older needles	3		1.5	0.80	0.424
Yellowing of current needles	0		-	-	-
Current yellowing type	0		-	-	-
Yellowing of older needles	0		-	-	-
Older yellowing type	0		-	-	-
Overall discoloration	1		-	-	-
Mechanical damage type	3		0	1.60	0.112
Mechanical damage %	3	1.54			0.127
Stem damage type	6		0	2.20	**0.028**
Stem damage %	7	2.27			**0.025**
Insect activity	9		22.0	0.06	0.952
Fungi	3		1.5	0.80	0.424

control observer failing to record a particular form of damage (extraction damage) on a number of occasions. This occurred when stems had been checked for damage earlier by the assessor.

Scots pine

The checks on the Scots pine assessments revealed a large number of differences. Scots pine is generally recognised as one of the most difficult species to assess, and the lack of consistency in the scoring confirms this. Significant differences in scoring were noted for canopy closure, defoliation type, shoot death location and extent, flowering, leader condition, branch density, browning of needles, overall discoloration, the extent of mechanical damage to the crown and the extent of shoot death.

Differences in defoliation type were widely distributed with no particular pattern being apparent. Many of the trees classed by surveyors as having defoliation of the lower crown were considered to have gap-like or uniform defoliation by the control observer. Flowering was generally scored higher by the surveyors than by the control observer. Confusion clearly existed over the definition of a 'normal' leader and a side shoot that had taken over as leader. Given time, a side shoot leader will become indistinguishable from a normal leader and this may have been the source of the discrepancy. The differences in browning scores were the result of higher scores being given by the surveyors than by the control observer for browning of current needles and the reverse for browning of older needles. Overall discoloration was scored higher by the surveyors.

Table 3c. Reliability of Scots pine data, as indicated by comparisons between the surveyors' and the check team's scores. Sample size: 198. Neither test could be used when the number of differences was less than three. Significant ($p < 0.05$) differences between the assessments are shown in bold.

Variable	Number of differences	t-test t	Wilcoxon test T	Z	p
Dominance	34		210.0	1.50	0.138
Canopy closure	66		185.5	5.88	**0.000**
Crown form	30		226.0	0.13	0.894
Defoliation type	56		420.0	3.08	**0.002**
Crown density	56	−0.23			0.818
Shoot death location	83		850.0	4.05	**0.000**
Shoot death extent	81		394.5	5.96	**0.000**
Flowering (top)	69		417.5	4.72	**0.000**
Flowering (bottom)	87		922.0	4.20	**0.000**
Leader condition	58		122.5	5.68	**0.000**
Coning	50		481.0	1.51	0.134
Branch density	94		192.5	7.69	**0.000**
Needle retention	24		120.0	0.86	0.393
Browning of current needles	12		12.0	2.12	**0.034**
Browning of older needles	7		0	2.36	**0.018**
Yellowing of current needles	2		-	-	-
Current yellowing type	3		0	1.60	0.112
Yellowing of older needles	10		20.0	0.76	0.446
Older yellowing type	14		30.0	1.41	0.160
Overall discoloration	17		17.0	2.82	**0.005**
Mechanical damage type	41		291.5	1.80	0.071
Mechanical damage %	46	2.69			**0.008**
Stem damage type	11		26.0	0.62	0.535
Stem damage %	25	−3.01			**0.003**
Insect activity	19		81.5	0.54	0.588
Fungi	0		-	-	-

Oak

There were also problems with the assessment of oak. Differences in scoring were identified for canopy closure, crown density, defoliation type, dieback location, the number of epicormic shoots, leaf browning type, acorns and insect activity.

The problem with crown density assessments was only apparent in oak. As it is one of the most important assessments, the differences have been looked at in more detail (Table 4). Generally, the scores made by the control observer were higher than those made by the individual surveyors. In one case, a difference of 20% was recorded. However, overall, differences were much less than noted in previous years.

Defoliation types were recorded differently, with the degree of severity of gaps in the crown being the main source of disagreement. Epicormics were generally scored as being less abundant by the surveyors, whereas they recorded acorns more frequently. Insect activity scores showed a wide variation and scoring clearly needs to be improved.

Beech

Relatively few differences were noted for beech. Canopy closure, crown form (Roloff scores), mechanical damage to the crown, premature leaf loss and insect activity all showed differences between the two visits. Problems have been encountered with crown form in the past, and the discrepancies, although significant, are much

Table 3d. Reliability of oak data, as indicated by comparisons between the surveyors' and the check team's scores. Samples size: 167. Neither test could be used when the number of differences was less than three. Significant ($p < 0.05$) differences between the assessments are shown in bold.

Variable	Number of differences	t-test t	Wilcoxon test T	Z	p
Dominance	10		16.5	1.12	0.265
Canopy closure	78		402.0	5.67	**0.000**
Crown density	50	−3.30			**0.001**
Defoliation type	47		315.0	2.63	**0.008**
Dieback type	32		193.0	1.33	0.187
Dieback location	68		851.0	1.97	**0.049**
Dieback extent	43	−0.20			0.844
Crown form	39		314.5	1.05	0.294
Stem epicormics	70		686.0	3.26	**0.001**
Branch epicormics	66		577.5	3.37	**0.001**
Browning of leaves	4		0	1.83	0.068
Browning type	16		16.0	2.69	**0.007**
Yellowing of leaves	1		-	-	-
Yellowing type	2		-	-	-
Overall discoloration	2		-	-	-
Acorns	13		6.0	2.76	**0.006**
Mechanical damage type	44		459.5	0.41	0.680
Mechanical damage %	47	0.292			0.771
Stem damage type	1		-	-	-
Stem damage %	2		-	-	-
Insect activity	87		777.0	4.81	**0.000**
Fungi	0		-	-	-

Table 4. Differences between the assessments of crown density of oak made by a control observer and individual surveyors. The numbers of trees in each crown density % category are shown. Identical assessments are portrayed in bold.

Control observer	Individual surveyor scores (%)													
	5	10	15	20	25	30	35	40	45	50	55	60	65	70
5	**0**	0	0	0	0	0	0	0	0	0	0	0	0	0
10	1	**1**	0	0	0	0	0	0	0	0	0	0	0	0
15	1	1	**6**	0	0	0	0	0	0	0	0	0	0	0
20	0	0	3	**16**	1	0	0	0	0	0	0	0	0	0
25	0	0	1	0	**11**	0	2	0	0	0	0	0	0	0
30	0	0	1	2	4	**18**	4	0	0	0	0	0	0	0
35	0	0	0	0	1	2	**23**	2	1	0	0	0	0	0
40	0	0	0	0	0	3	4	**19**	4	0	0	0	0	0
45	0	0	0	0	1	0	1	1	**7**	1	0	0	0	0
50	0	0	0	0	0	0	2	1	5	**0**	0	0	0	0
55	0	0	0	0	0	0	0	0	1	4	**0**	0	0	0
60	0	0	0	0	0	0	0	1	0	0	1	**0**	0	0
65	0	0	0	0	0	0	0	0	0	1	2	**2**	0	0
70	0	0	0	0	0	0	0	0	0	0	0	0	0	**4**

lower than in previous years, with only 5 out of 88 trees having different scores. Premature leaf loss also involved a relatively small number of trees and the difference was due to one of the surveyors not scoring this variable when it was present in very small quantities.

General conclusions

It is clear that there are a number of problems associated with the consistency of the observations, despite the attention that has been paid to this aspect of the programme (Innes, 1990). Crown density estimates in 1990 were much more consistent than in previous years, reflecting the emphasis that has been put on this variable in training; 95% of observations can now be expected to be within 10% of the control observer's score, compared to 95% being within 15% to 20% in previous years. Other indices were scored less consistently. In some cases, this appears to be the result of misinterpretation of

Table 3e. Reliability of beech data, as indicated by comparisons between the surveyors' and the check team's scores. Sample size: 88. Neither test could be used when the number of differences was less than three. Significant ($p < 0.05$) differences between the assessments are shown in bold.

Variable	Number of differences	t-test t	Wilcoxon test T	Z	p
Dominance	2	-	-	-	-
Canopy closure	25		82.5	2.15	**0.031**
Crown density	12	−0.575			0.567
Defoliation type	12		27.5	0.90	0.339
Dieback type	14		49.0	0.22	0.827
Dieback location	19		82.5	0.50	0.616
Dieback extent	38	0.992			0.324
Crown form	5		0	2.02	**0.043**
Leaf size	5		3.0	1.21	0.227
Frequency of rolling	13		23.0	1.57	0.119
Degree of rolling	14		28.0	1.54	0.127
Leaf browning	8		8.0	1.40	0.164
Browning type	9		21.0	0.18	0.858
Leaf yellowing	7		12.0	0.34	0.736
Yellowing type	4		2.0	1.09	0.276
Overall discoloration	1		-	-	-
Mast	20		62.5	1.59	0.115
Mechanical damage type	24		0	4.29	**0.000**
Mechanical damage %	15	2.86			**0.005**
Stem damage type	8		12.5	0.77	0.443
Stem damage %	8	0.64			0.525
Premature leaf loss	8		0	2.52	**0.012**
Insect activity	35		18.0	4.86	**0.000**
Fungi	0		-	-	-

the instructions, in others, observer error may be involved. The reduction from two surveyors per team to one that occurred in 1990 is a further factor which undoubtedly affected the reliability of the results. Further analysis is being undertaken to assess whether the differences are widespread amongst surveyors or whether they are due to one or two individuals.

Changes during the assessment period

The 1990 programme was undertaken over the period 10 July to 31 August. During this time there was little, if any, rainfall over much of England. For example, July and August rainfall figures for Bournemouth were 14.6 mm and 27.3 mm and for Sheffield were 21.7 mm and 38.2 mm, respectively. In Scotland, rainfall was considerably higher, with the July and August figures for Aviemore being 30.7 mm and 88.6 mm and for Rothesay on the Isle of Bute being 86.5 mm and 87.4 mm, respectively. Consequently, in some areas, trees progressively showed increased signs of drought stress. This represents a major problem for a study of this nature as trees assessed at the beginning of the period were unlikely to be affected whereas those assessed at the end of the period might well be expected to show symptoms associated with drought.

The extent of the effect was determined by repeating assessments made on some of the plots visited near the beginning of the period. Seven plots each of Norway spruce, Scots pine and beech were re-assessed, with the interval between sampling being approximately 7 weeks. The results are difficult to interpret as any difference in score is likely to be due to a real change, combined with differences associated with observer variation (attributable to, for example, differences in the light conditions between the two assessments). Although the latter was minimised by the use of the same surveyor for each pair of plots, small changes between the two visits are likely to be of little relevance unless there is definite evidence of trend. The differences associated with the observations should be randomly distributed and, consequently, any directional change, indicated by a significant difference between the pairs of observations, should reflect a real change in condition.

Changes have been examined using t-tests for paired samples. The results are given in Table 5. Many of the indices were significantly different, suggesting that changes had occurred. In all cases, the changes were in the expected direction.

Norway spruce

Crown density scores in the repeat assessments were generally higher than on the first assessment, indicating that crown thinning had occurred (Table 6a). In some cases, changes were substantial, amounting to 20% to 25% of the density. 23 trees showed an improvement during the period, with the majority involving an improvement of 5% (one class). This is within the confidence limit for the assessments (see next section), and ± one class probably has relatively little meaning. However, the substantial deterioration of some trees is likely to be real.

The extent of shoot death in the crown also increased during the course of the programme. As might be expected, very few trees (1) showed less shoot death on the repeat visit. Eleven (7%) of trees had a greater amount of shoot death. Of these, nine were trees where shoot death was first classified as scattered or common, whereas only one of 69 trees with shoot death classified as absent on the first visit had shoot death recorded during the repeat assessment.

Needle retention was significantly lower on the second visit, although this is an index that is known to be subject to error (Innes and Boswell, 1989), as illustrated by the nine trees which had an apparent increase in needle retention (from 7 years to more than 7 years). The results are shown in Table 6b. Particular notice should be taken of the tree scored as having 5 years of needles on the first visit and only one year on the repeat assessment and the 11 trees scored as having 7 years of needles on the first visit and five or less on the second. The directional changes are consistent with the changes in crown density. A comparison of the trees that changed density and those that changed needle retention reveals some inconsistencies (Table 6c). However, a direct relationship between crown density and needle retention is not always apparent as the most common type of crown thinning experienced in Britain involves the uniform loss of needles throughout the crown rather than from the oldest needles to the youngest.

Needle discoloration showed little change, although there was a significant increase in the amount of yellowing of current needles and overall discoloration. Eleven trees with less than 10% of current needles showing yellowing at the beginning of the programme had more

Table 5. Changes in scores for various indices; t-values are given. A negative value indicates that the scores for a particular variable tended to be higher on the repeat assessment. Significant ($p < 0.05$) changes are shown in bold type.

Variable	Norway spruce	Scots pine	Beech
Crown density	**−6.06**	−1.65	**−4.93**
Shoot death extent	**−2.95**	−1.51	
Needle retention	**5.27**	**2.90**	
Current needle browning	−1.74	**−3.67**	
Older needle browning	1.00	**−4.17**	
Current needle yellowing	**−2.39**	−1.42	
Older needle yellowing	−1.00	−1.91	
Leaf browning			**−2.61**
Leaf yellowing			**−6.58**
Overall discoloration	**−2.36**	**−5.39**	**−7.25**
Dieback			**−2.23**
Leaf size			0.82
Leaf-rolling frequency			**−4.98**
Leaf-rolling degree			**−4.55**
Premature leaf loss			**−2.58**

(Significance levels: $p = 0.05$: 1.98; $p = 0.01$: 2.62)

Table 6a. Changes in the crown condition of Norway spruce between the beginning and end of the survey. Values are the number of trees in each category.

Assessment at start	Change in crown density (%)								
	−15	−10	−5	None	+5	+10	+15	+20	+25
0	0	0	0	4	2	0	0	0	0
5	0	0	1	6	1	1	0	1	0
10	0	0	1	3	8	1	1	0	0
15	0	0	2	4	4	3	1	0	1
20	0	0	2	12	9	2	1	0	0
25	0	0	1	12	7	1	1	1	0
30	0	0	4	11	5	2	0	0	0
35	0	0	1	7	2	7	0	0	0
40	0	0	4	4	1	0	0	1	0
45	0	1	2	1	4	1	0	0	0
50	0	1	1	1	2	0	0	1	0
55	0	0	0	0	1	0	0	0	0
60	0	0	0	1	0	0	0	0	0
65	0	0	0	2	0	1	0	0	0
70	0	1	1	0	0	0	0	0	0
75	0	0	0	0	0	0	0	0	0
80	0	0	0	1	0	0	0	0	0
85	0	0	0	0	1	0	0	0	0
90	0	0	0	0	0	0	0	0	0
95	0	0	0	1	0	0	0	0	0

Table 6b. Changes in the needle retention in Norway spruce during the programme. Values are the number of trees in each category.

Repeat assessment	Needle retention on first visit								
	0	1	2	3	4	5	6	7	>7
0	0	0	0	0	0	0	0	0	
1	0	2	0	0	0	1	0	0	0
2	0	0	0	0	0	0	0	0	0
3	0	0	0	1	2	1	0	0	0
4	0	0	0	0	4	3	3	2	0
5	0	0	0	0	0	22	9	9	0
6	0	0	0	0	0	0	28	6	0
7	0	0	0	0	0	0	0	18	7
>7	0	0	0	0	0	0	0	9	41

Table 6c. Comparison of changes of needle retention with changes in crown density for Norway spruce. A score of 5 for crown density indicates that the crown density decreased by 5% during the period (i.e. the score increased) and a score of −1 for needle retention indicates that needle retention decreased by one year during the period. Values are the number of trees in each category.

Needle retention difference	Crown density difference							
	25	20	15	10	5	0	−5	−10
1	0	0	0	0	4	5	0	0
0	0	2	1	9	31	56	14	3
−1	0	0	0	6	8	7	6	0
−2	1	1	3	3	3	2	0	0
−3	0	0	0	1	1	0	0	0
−4	0	1	0	0	0	0	0	0

yellowing at the end, with the amount of yellowing increasing to 25–60% in five of the trees.

Scots pine

Generally, there was little change in crown density during the programme. Only one tree had an assessment differing by more than 10% and, as this was in a positive (improving) direction, it is unlikely to be real. The extent of shoot death was also unaffected.

Needle retention was significantly lower on the second visit. The change mainly involved the trees scored as having three years of needles on the first occasion, with 15 (out of 88) of these having two years' retention on the second visit.

There was an increase in the amount of browning, both of current and of older needles. In both cases, this increase primarily involved trees that had been scored as having less than 10% browning on the first visit. The changes resulted in an increase in the number of trees scored for overall discoloration.

Beech

Of the three species investigated, beech showed the greatest changes in crown condition during the period of the programme. All the main indices changed significantly between the two assessments. Crown thinning occurred, with individual trees losing up to 45% of their crown density between assessments. A number of trees showed minor improvements, but it is not clear whether this represents the greater error surrounding the assessment of beech or a genuine improvement caused by a late flush of leaves. Of the two, the former is more likely as the drought would have reduced or prevented any late flushes. This is supported by the absence of any change in the frequency of small leaves.

Discoloration was significantly worse on the second visit, with both browning and yellowing increasing. The increase in yellowing (Table 6e) was much more apparent than the increase in browning, which mainly involved 12 (out of 144) trees scored with less than 10% browning on the first visit. In the case of yellowing, the changes again primarily involved trees scored with less than 10% yellowing on the first visit, with 43 (out of 141) showing an increase. Overall discoloration was similarly affected, with 35% of the trees scored as having less than 10% discoloration on the first visit showing discoloration on the second visit.

Premature leaf loss, indicated by green leaves on the ground, was not recorded on the first visit. However, on the second visit, it was recorded as infrequent under 29 trees, common under seven and abundant under one.

The degree and extent of leaf-rolling both sig-

Table 6d. Changes in the crown density scores of Scots pine during the programme. Values are the numbers of trees in each category.

Assessment at start	Change in crown density (%)						
	−20	−15	−10	−5	None	+5	+10
0	0	0	0	0	3	1	1
5	0	0	0	0	4	3	1
10	0	0	0	0	11	5	1
15	0	0	0	1	5	11	0
20	0	0	1	1	12	7	5
25	0	0	1	6	16	6	1
30	0	0	1	5	17	8	0
35	0	0	0	6	7	2	0
40	0	0	1	1	2	0	0
45	0	1	1	0	2	0	0
50	1	0	0	2	0	0	1
55	0	0	0	0	2	0	0
60	0	0	0	0	0	0	0
65	0	0	0	0	0	0	0
70	0	0	0	0	0	0	0
75	0	0	0	0	0	0	0
80	0	0	0	0	0	0	0
85	0	0	0	0	0	0	0
90	0	0	0	0	0	0	0
95	0	0	0	0	0	0	0

Table 6e. Changes in the extent of crown chlorosis (yellowing) in beech during the programme. Values are the numbers of trees in each category.

	First visit			
Second visit	0-10%	11-25%	26-60%	>60%
0–10%	98	0	0	0
11–25%	29	8	0	0
26–60%	14	0	2	0
>60%	0	2	0	5

Table 6f. Changes in the extent of leaf-rolling in beech during the survey. 0: absent; 1: leaves on a few shoots in the upper crown; 2: leaves on a few shoots elsewhere; 3: leaves on about 50% of shoots in upper crown; 4: leaves on about 50% of shoots throughout crown; 5: leaves on most shoots in upper crown; 6: leaves on most shoots throughout the crown; 7: virtually all leaves affected. Values are the numbers of trees in each category.

Second visit	First visit							
	0	1	2	3	4	5	6	7
0	**56**	1	0	1	0	1	1	0
1	20	**9**	0	0	0	0	0	0
2	1	0	**0**	0	0	0	0	0
3	8	2	0	**10**	0	0	0	0
4	4	0	1	0	**2**	1	0	0
5	5	3	0	4	0	**23**	0	0
6	1	1	1	1	2	0	**5**	0
7	0	1	0	1	0	0	2	**0**

nificantly increased during the programme. This is as expected, as leaf-rolling in broadleaves is a symptom of drought (Bernier et al., 1989; Skelly et al., 1987). The increase primarily involved trees scored as without rolling on the first visit. On trees where rolling was recorded on the first visit, the degree of rolling showed little change. However, the extent of rolling generally increased, whatever the extent on the first visit (Table 6f).

An increase in the amount of dieback was also recorded. This is difficult to explain as dieback is normally regarded as a long-term process. However, observations suggest that the loss of shoots and small branches may occur quite quickly (Figure 2). The main change was from 'leaf loss only' to 'dieback restricted to small branches', suggesting the loss of first-order, leafless shoots during the period. There is also the possibility that the premature dropping of leaves led to the upper crown being more visible and dieback missed on the first occasion was seen on the second.

Spatial patterns of changes

The locations of the plots used for the reassessments are show in Figure 3. Plots were chosen from throughout the country in order to assess spatial differences in the extent of the changes. Only seven plots of each species were examined and the results presented below must therefore be viewed with caution.

The intensity of the drought at each site has been assessed in terms of the soil moisture deficit. Although these figures refer to agricultural soils, they provide an indication of the relative severity of drought in different parts of the country. The results are presented in Table 6h.

Substantial differences exist between sites. For Norway spruce, the greatest changes in

Table 6g. Changes in the crown density scores for beech during the programme. Values are the number of trees in each category.

Assessment at start	Change in crown density (%)											
	−10	−5	None	+5	+10	+15	+20	+25	+30	+35	+40	+45
0	0	0	2	0	0	0	0	0	0	0	0	0
5	0	0	4	2	0	0	0	0	0	0	0	0
10	0	0	6	3	2	1	0	0	0	0	0	0
15	0	2	5	1	1	0	0	0	0	0	0	0
20	0	1	10	7	4	0	0	1	0	0	0	0
25	0	5	6	7	4	1	0	0	0	0	0	0
30	0	1	13	6	4	0	0	0	0	0	0	0
35	2	11	14	4	2	0	0	0	0	0	0	1
40	0	2	5	3	4	0	0	0	0	0	0	0
45	0	5	2	2	1	0	0	0	0	0	0	0
50	0	1	2	0	0	0	0	0	1	0	0	0
55	0	1	0	1	0	0	0	0	0	0	0	0
60	0	1	0	0	0	0	0	0	0	0	0	0
65	0	0	1	0	1	0	0	0	0	0	0	0

Figure 2. Loss of minor branches in an oak during the period 1987–90.

crown density and needle retention occurred at a site with relatively little drought stress. It seems likely that some other (unknown) factor was involved. All three sites with low drought stress showed no changes in needle discoloration, contrasting markedly with sites experiencing relatively high drought stress. Crown densities decreased at all sites, regardless of the degree of drought stress. The changes in discoloration therefore appear to be attributable to drought, whereas the needle losses occurred as part of the normal changes during the year.

A similar situation exists for Scots pine, with changes in crown density, needle retention and the extent of shoot death appearing to be independent of the degree of drought stress. Sites experiencing drought stress had an increase in the amount of current needle browning and overall discoloration; no such pattern was apparent in sites without drought. The cause of the increase in older needle browning at one of the low drought stress sites is unknown.

In beech, changes in crown density were also independent of the degree of drought stress. One of the sites (no. 17), located in north-west Wales, appears to have a number of anomalous changes. The reasons for these are unknown, but are probably site-related. Changes in the amount of rolling (both frequency and extent)

Table 6h. Changes in crown condition in 1990 relative to drought intensity. 125.0 is the highest soil moisture deficit that can be attained and indicates that no soil moisture was available (to grass). Values for individual indices are the sums of the indices for individual trees. With the exception of needle retention, a minus sign indicates that the relevant index was higher (worse) on the second visit. A lower (i.e. worse) value for needle retention on the second visit produced a positive value.

	Site						
Norway spruce	1	2	3	4	5	6	7
Soil moisture deficit 10 July 1990	118.5	69.1	101.4	2.4	120.6	7.5	8.4
Soil moisture deficit 28 August 1990	125.0	108.2	125.0	13.4	125.0	56.1	26.5
Crown density %	−80	−20	−35	−50	−50	−30	−195
Needle retention	+6	+6	0	+5	−8	0	+44
Shoot death extent	−2	−1	0	0	0	0	−7
Current needle browning	−3	0	0	0	0	0	0
Older needle browning	0	0	+1	0	0	0	0
Current needle yellowing	+3	−2	0	0	−13	0	0
Older needle yellowing	0	0	0	0	−1	0	0
Overall discoloration	−1	0	0	0	−8	0	0

Scots pine	8	9	10	11	12	13	14
Soil moisture deficit 10 July 1990	118.5	59.3	7.5	0.4	8.4	69.1	119.2
Soil moisture deficit 28 August 1990	125.0	109.5	56.1	13.4	26.5	108.2	125.0
Crown density %	−10	−40	−20	−40	−70	−35	−55
Needle retention	+2	+3	0	+7	−2	0	+6
Shoot death extent	−1	−1	0	0	−3	0	−1
Current needle browning	−2	−2	0	0	0	−3	−11
Older needle browning	−5	0	−10	0	0	0	−3
Current needle yellowing	0	0	0	0	0	0	−2
Older needle yellowing	0	−8	0	0	0	0	+2
Overall discoloration	−7	−8	0	0	0	−4	−12

Beech	15	16	17	18	19	20	21
Soil moisture deficit 10 July 1990	120.6	112.7	8.4	118.5	30.4	25.5	68.6
Soil moisture deficit 28 August 1990	125.0	125.0	26.5	125.0	74.4	62.6	112.8
Crown density %	−80	−95	−125	−105	−10	−15	+35
Crown pattern	−8	−9	−24	−27	+2	−6	−3
Dieback extent	−30	−19	−25	−10	0	−10	−22
Rolling frequency	−48	−43	−25	+6	−1	−12	−38
Rolling degree	−3	−17	+1	+1	−1	−8	−17
Leaf browning	−6	−8	0	−2	0	+2	−1
Leaf yellowing	−24	−8	0	−18	0	0	−11
Overall discoloration	−30	−18	0	−20	0	+1	−11
Premature leaf loss	−1	−7	0	0	0	0	0

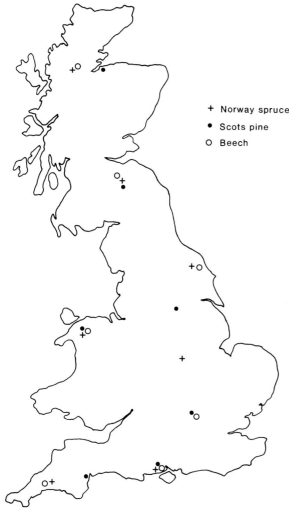

Figure 3. Location of plots examined at both the start and the end of the survey period.

and discoloration (yellowing and overall) were higher at drought-stressed sites. Leaf browning also increased at two of the four drought-stressed sites.

The main effect of the drought conditions in 1990 appears to have been an increase in the amount of leaf and needle discoloration. Although changes in crown density were recorded, these cannot be readily related to the degree of drought stress. In many cases, the changes were significant (Table 6h), suggesting that great care is required in interpreting the results for 1990, both in terms of comparisons with previous years and comparisons between sites.

Results

In the following section, two types of results are described. The first concerns the results of assessments made in 1990 and refers to all trees. The 1990 results have also been used to draw comparison with results from previous years. However, in the latter case, only trees common to the comparison period have been used. Consequently, the sample sizes for these common sample trees (the CST sample) are considerably smaller than the sample size for 1990 alone (Table 7). For example, only 40% of the oak trees assessed in the period 1987 to 1990 were assessed in all four years, whereas 96% of the sample was assessed in 1990.

In all the tables in the 'Results' section, figures have been rounded to either no or one decimal place. Consequently, percentages may not always total to 100.

Crown density

Crown density was assessed in 5% classes, following the recommendations of the Commission of the European Communities and the International Co-operative Programme on Monitoring and Assessment of Air Pollution Effects on Forests of the United Nations Economic Commission for Europe. The overall data for crown density are presented in Table 8a and Figure 4.

Table 7. Sample sizes of each species used in constructing Table 8. The percentage of trees assessed relative to the total number of trees of each species involved in the programme is given.

	Sitka spruce	Norway spruce	Scots pine	Oak	Beech
Total number of trees assessed 1987–90	1626	1996	2101	1820	905
Number of trees assessed in 1990	1344 83%	1752 88%	1944 93%	1740 96%	864 95%
Number of common sample trees 1987–90	1064 65%	1431 72%	1348 64%	724 40%	642 71%

Table 8a. The percentage of trees of each species in each crown foliage density category. A score of zero indicates no loss of density. A score of 5% represents a loss in crown density of 1–5% and so on.

Score	Sitka spruce	Norway spruce	Scots pine	Oak	Beech
0	1.9	5.8	2.2	1.7	3.6
5	3.2	6.7	3.4	4.0	4.4
10	6.0	9.6	7.4	7.5	8.6
15	13.5	13.9	11.8	9.5	8.6
20	12.9	14.2	17.2	11.6	12.2
25	13.8	13.5	17.2	11.4	13.3
30	12.1	11.9	14.9	13.3	13.0
35	10.9	9.8	10.3	14.9	12.3
40	8.0	5.4	6.6	9.9	7.8
45	6.2	3.0	2.9	5.4	6.3
50	3.1	2.3	1.6	3.5	4.5
55	2.1	1.1	0.8	2.2	2.9
60	2.1	0.5	1.0	1.6	0.6
65	1.9	1.0	0.2	1.2	1.0
70	0.7	0.4	0.3	1.0	0.2
75	0.6	0.1	0.3	0.5	0.1
80	0.9	0.2	0.7	0.5	0.2
85	0.1	0.3	0.2	0.4	0.3
90	0	0.1	0.1	0	0.1
95	0	0.1	0.3	0.1	0.1
Dead	0	0.1	0.7	0	0

Comparison between years can be made using the CST sample. The comparisons are given in Table 8b and Figure 5.

Sitka spruce

The crown density of Sitka spruce generally improved in 1990. This was mainly the result of increased needle mass following the severe defoliation caused by the green spruce aphid in 1988–89. Some trees assessed in 1989 had only one year's needles in the crowns and, by 1990, these mostly had two years of needles (see below). Trees with two or three years' needles in 1989 also showed a tendency for increased needle longevity. The improvements were particularly apparent amongst the more severely defoliated trees, confounding views that severely affected conifers are unlikely to recover (e.g. Jukola-Sulonen *et al.*, 1987).

Over the 4-year period, the crown density scores for Sitka spruce have fluctuated considerably. Densities decreased between 1987 and 1988 and, in 1989, some trees improved while others deteriorated as a result of the aphid defoliation. As mentioned above, substantial improvements in crown density occurred in 1990.

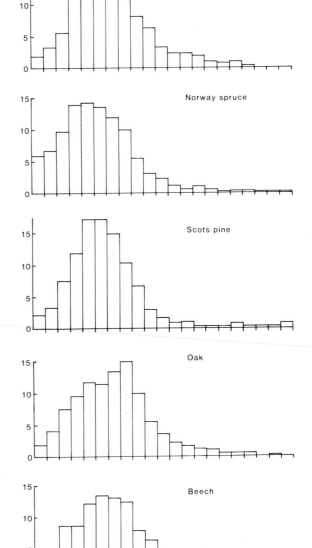

Figure 4. Distribution of crown density classes for each species in 1990. Scores are given in 5% classes, with 0 indicating no reduction in density, 5% representing 1–5%, 10% representing 6–10% and so on.

Table 8b. Comparison of tree crown density results for the period 1987–1990. Ten per cent categories have been used as the 5% class interval was only introduced in 1989. Class 0 represents a 0–10% reduction in density, class 1 represents 11–20% reduction, class 2 represents 21–30% reduction and so on. All figures are given in percentages of trees in each class in each year (i.e. rows sum to 100).

	Crown density class									
	0 0-10%	1 11-20%	2 21-30%	3 31-40%	4 41-50%	5 51-60%	6 61-70%	7 71-80%	8 81-90%	9 91-100%
Sitka spruce										
1987	15.4	22.8	24.1	18.8	11.3	5.2	2.2	0.2	0.1	0
1988	8.0	23.5	26.0	22.3	13.6	5.2	1.1	0.3	0	0
1989	11.2	22.8	24.3	19.7	10.2	4.7	2.9	2.2	1.6	0.5
1990	17.3	24.7	23.0	17.8	7.2	4.0	2.3	2.1	1.3	0.3
Norway spruce										
1987	21.2	23.7	24.1	14.8	8.9	4.4	2.0	0.6	0.3	0
1988	20.2	26.2	24.2	16.6	7.3	3.6	1.2	0.3	0.3	0
1989	25.2	27.6	22.2	14.4	6.2	2.4	1.3	0.6	0.3	0
1990	20.9	28.6	25.2	15.4	5.8	1.7	1.4	0.3	0.5	0.1
Scots pine										
1987	17.8	23.1	24.9	19.1	8.2	3.2	1.0	1.3	0.9	0.6
1988	6.6	22.2	32.1	20.8	10.1	3.6	1.4	1.3	0.9	1.0
1989	10.9	29.7	28.0	16.6	7.5	3.1	1.9	1.2	0.6	1.0
1990	9.9	28.4	33.5	17.2	5.1	2.2	0.7	1.3	0.4	1.3
Oak										
1987	8.1	14.8	22.0	29.6	14.2	5.7	3.9	1.5	0.3	0
1988	4.1	16.2	29.7	28.7	12.6	5.7	1.9	0.6	0.6	0
1989	6.4	20.8	30.4	26.6	10.4	3.2	1.3	0.4	0.5	0
1990	9.5	24.3	32.2	21.7	8.0	2.6	1.0	0.3	0.4	0
Beech										
1987	8.3	22.6	26.2	27.0	12.5	2.7	0.8	0	0	0
1988	9.7	24.2	32.4	22.5	8.9	1.6	0.5	0.3	0	0
1989	18.6	30.5	28.9	16.1	4.2	1.1	0.6	0	0	0
1990	17.0	20.9	26.0	19.7	10.8	3.3	1.3	0.4	0.5	0.1

Following individual trees through time can produce important information on changes in forest condition (Innes and Boswell, 1990b), although a number of uncertainties are involved (Mahrer, 1989). Since 1987, many of the trees have changed substantially (Table 8c). For example, two trees scored in class 1 (11–20% loss of density) in 1987 were scored in class 7 (70–80% loss of density) in 1990. Substantial improvements are also apparent. Most trees have shown some change during the period, with the majority of trees changing by at least one class; the figures shown in Table 8c actually underestimate the numbers of trees that have changed as no account is taken of those trees that changed in 1988 or 1989 and returned to their 1987 score in 1990. As percentages have been used, the values in the higher scores should be interpreted with caution as the sample sizes are very small.

Norway spruce
The crown density of Norway spruce has remained very similar during the 4-year period. Changes in the common sample trees between 1987 and 1989 are shown in Table 8d. Percentages of trees with the same score in 1987 and 1990 are higher than for Sitka spruce, indicating greater stability.

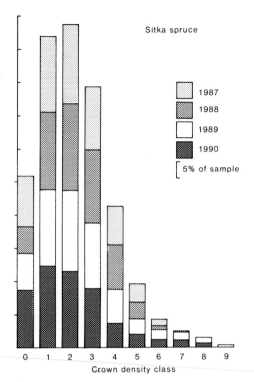

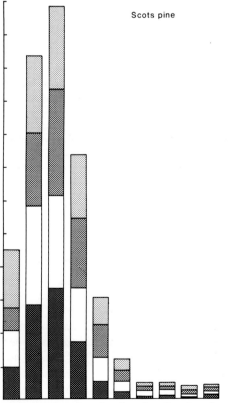

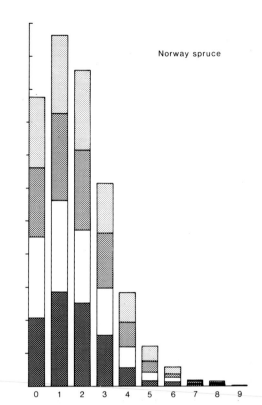

Figure 5. Distribution of crown density scores for common sample trees over the period 1987–90. A score of 0 indicates a reduction in density of 0–10%, 1 represents 11–20%, 2 represents 21-30% and so on.

Scots pine

Scots pine also showed little change in comparison to 1989 although there have been considerable changes to many of the trees during the period 1987–90 (Table 8e). In 1990, a small decrease in the number of the trees with 40–80% density loss was recorded, although the number of dead trees increased. Since 1987, there has been a general improvement in condition, although a number of individuals have deteriorated markedly, with trees scored in class 1 in 1987 being scored in classes 8 and 9 in 1990. However, the majority of trees currently in classes 8 and 9 in 1990 were in classes 7, 8 or 9 in 1987.

Oak

The improvement in oak that occurred between 1988 and 1989 continued into 1990. The propor-

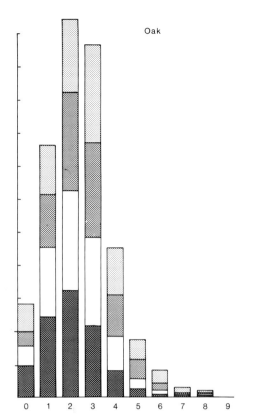

Oak

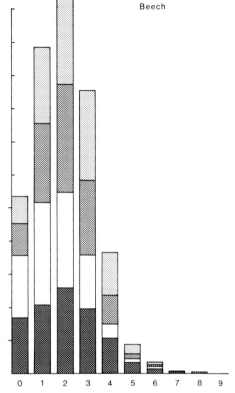

Beech

tion of sparsely foliaged trees is continuing to decline, as is clearly shown in Table 8f. As with the other species, the greatest changes have occurred amongst the more severely defoliated trees, with the majority improving.

Beech

Some of the most marked changes between 1989 and 1990 occurred in beech. The proportion of trees in the most dense category increased substantially, but this was matched by

Table 8c. Changes in the crown density of individual Sitka spruce trees between 1987 and 1990. Values represent the percentage of trees in a given class in 1987 that were in the specified class in 1990 (i.e. columns sum to 100).

	Crown density class in 1987									
	0 0-10%	1 11-20%	2 21-30%	3 31-40%	4 41-50%	5 51-60%	6 61-70%	7 71-80%	8 81-90%	9 91-100%
Crown density class in 1990 (n = 1064)										
0	**26.2**	16.5	10.2	3.0	0.8	0	0	0	0	0
1	37.8	**31.3**	30.5	19.0	8.3	7.3	0	0	0	0
2	22.6	28.4	**27.3**	28.5	20.8	21.8	8.7	0	0	0
3	12.2	16.5	18.4	**25.5**	29.2	18.2	13.0	0	0	0
4	0.6	5.3	6.3	14.5	**20.0**	20.0	34.8	0	0	0
5	0.6	0.8	6.6	4.0	5.8	**10.9**	17.4	0	0	0
6	0	0.4	0.4	4.5	8.3	12.7	**8.7**	100.0	100.0	0
7	0	0.8	0.4	1.0	5.8	9.1	13.0	**0**	0	0
8	0	0	0	0	0.8	0	4.3	0	**0**	0
9	0	0	0	0	0	0	0	0	0	**0**

Table 8d. Changes in the crown density of Norway spruce between 1987 and 1990. Values represent the percentage of trees in a given class in 1987 that were in the specified class in 1990 (i.e. columns sum to 100).

	Crown density class in 1987									
	0	1	2	3	4	5	6	7	8	9
	0-10%	11-20%	21-30%	31-40%	41-50%	51-60%	61-70%	71-80%	81-90%	91-100%
Crown density class in 1990 (n = 1431)										
0	**46.5**	24.5	15.9	8.0	2.3	0	0	0	0	0
1	35.6	**42.5**	31.9	15.6	10.2	1.6	0	0	0	0
2	15.2	26.5	**33.6**	34.4	21.1	7.9	10.7	11.1	0	0
3	2.6	5.6	14.2	**32.5**	42.2	31.7	7.1	0	0	0
4	0	0.6	3.8	8.5	**18.8**	30.2	25.0	0	0	0
5	0	0.3	0.6	0.9	3.1	**17.5**	14.3	11.1	0	0
6	0	0	0	0	1.6	7.9	**32.1**	33.3	25.0	0
7	0	0	0	0	0	1.6	10.7	**11.1**	0	0
8	0	0	0	0	0.8	0	0	33.3	**75.0**	0
9	0	0	0	0	0	1.6	0	0	0	**0**

Table 8e. Changes in the crown density of individual Scots pine between 1987 and 1990. Values represent the percentage of trees in a given class in 1987 that were in the specified class in 1990 (i.e. columns sum to 100).

	Crown density class in 1987									
	0	1	2	3	4	5	6	7	8	9
	0-10%	11-20%	21-30%	31-40%	41-50%	51-60%	61-70%	71-80%	81-90%	91-100%
Crown density class in 1990 (n = 1348)										
0	**22.9**	12.9	7.7	3.9	0	2.3	0	5.9	0	0
1	35.0	**36.7**	33.0	21.4	12.6	4.7	15.4	5.9	0	0
2	27.5	35.0	**37.2**	42.0	29.7	18.6	15.4	5.9	0	0
3	10.4	11.3	17.3	**24.9**	33.3	30.2	0	0	0	0
4	3.3	1.9	3.3	5.8	**10.8**	27.9	30.8	5.9	0	0
5	0.8	0.6	0.9	1.6	9.0	**9.3**	15.4	11.8	8.3	0
6	0	0.3	0	0	0.9	4.7	**7.7**	17.6	8.3	0
7	0	0.6	0.3	0	1.8	2.3	15.4	**29.4**	41.7	0
8	0	0.3	0	0.4	0	0	0	0	**16.7**	12.5
9	0	0.3	0.3	0	1.8	0	0	17.6	25.0	**87.5**

increase in the proportion of sparsely foliaged trees. In 1987, only one (out of 672) beech tree was scored as having lost more than 60% of its foliage. This tree was subsequently blown over and has been replaced, and it has therefore been omitted from the CST sample. Reference to Table 8g suggests that the crown density in 1987 is a poor indicator of the crown density in 1990, although the two values are significantly correlated (r = 0.52, $p > 0.001$).

Crown form and defoliation type

As in 1989, both crown form and defoliation type were assessed in all trees. As the samples have changed, particularly for oak, the results are presented below.

In the two spruce species, the most common form of defoliation involved the uniform loss of needles from throughout the crown. Loss of needles from the base of the crown upwards was also apparent. These two forms of defoliation

Table 8f. Changes in crown density of individual oak trees between 1987 and 1990. Values represent the percentage of trees in a given class in 1987 that were in the specified class in 1990 (i.e. columns sum to 100).

	Crown density class in 1987									
	0	1	2	3	4	5	6	7	8	9
	0-10%	11-20%	21-30%	31-40%	41-50%	51-60%	61-70%	71-80%	81-90%	91-100%
Crown density class in 1990 (n = 724)										
0	**44.1**	22.4	8.8	2.8	1.9	0	0	0	0	0
1	27.1	**35.5**	32.1	14.0	3.9	4.9	0	9.1	0	0
2	15.3	19.6	**32.1**	35.0	23.3	14.6	17.9	18.2	0	0
3	11.9	16.8	17.6	**32.2**	39.8	41.5	10.7	26.4	100.0	0
4	1.7	1.9	3.1	8.4	**19.4**	22.0	35.7	18.2	0	0
5	0	1.9	2.5	1.9	5.8	**4.9**	21.4	0	0	0
6	0	1.9	2.5	3.3	2.9	9.8	**7.1**	9.1	0	0
7	0	0	0	1.4	1.9	2.4	3.6	**9.1**	0	0
8	0	0	1.3	0.5	1.0	0	3.6	0	**0**	0
9	0	0	0	0.5	0	0	0	0	0	**0**

Table 8g. Changes in crown density of individual beech trees between 1987 and 1990. Values represent the percentage of trees in a given class in 1987 that were in the specified class in 1990 (i.e. columns sum to 100).

	Crown density class in 1987									
	0	1	2	3	4	5	6	7	8	9
	0-10%	11-20%	21-30%	31-40%	41-50%	51-60%	61-70%	71-80%	81-90%	91-100%
Crown density class in 1990 (n = 642)										
0	**41.9**	14.8	5.2	2.5	5.9	0	0	0	0	0
1	30.8	**24.3**	18.5	8.8	5.9	0	0	0	0	0
2	19.2	32.5	**27.2**	30.0	5.9	40.0	0	0	0	0
3	6.1	18.9	20.2	**32.5**	47.1	40.0	0	0	0	0
4	1.5	5.9	20.8	13.8	**17.6**	20.0	0	0	0	0
5	0.5	1.8	5.2	7.5	0	**0**	0	0	0	0
6	0	0.6	2.3	3.8	5.9	0	**0**	0	0	0
7	0	0.6	0	0	5.9	0	0	**0**	0	0
8	0	0.6	0.6	1.3	0	0	0	0	**0**	0
9	0	0	0	0	5.9	0	0	0	0	**0**

were more apparent in 1990 than in 1989. In Scots pine, the most common form of defoliation was also the uniform loss of needles from throughout the crown, with defoliation of the lower crown and gap-like defoliation also being important. The amount of trees classified with gap-like defoliation in 1990 increased; many of these were trees that had previously been classified as having no loss of foliage or loss from the base upwards.

The most common forms of defoliation in oak and beech involved either small gaps in the crown or gaps to the lateral branch systems. As might be expected, the values obtained in 1990 were almost identical to those obtained in 1989.

There has been some debate over the use of different indices of crown form. One type, favoured in the Federal Republic of Germany, involves the recognition of four stages (Roloff, 1985a, 1985b; Dobler *et al.*, 1988), although this

Table 9 (a). Defoliation types in spruce, 1990. 0: no obvious defoliation; 1: small window in upper crown; 2: large window; 3: top-dying; 4: uniform loss of needles throughout the crown; 5: peripheral defoliation; 6: loss of needles from base upwards; 7: other. Figures are percentages of trees in each species in each class (i.e. rows sum to 100).

	Defoliation type							
	0	1	2	3	4	5	6	7
Sitka spruce	25	1	1	0	51	0	18	4
Norway spruce	39	1	3	1	39	1	14	2

(b). Percentages of Scots pine with different defoliation types. 0: no obvious defoliation; 1: lower crown only; 2: peripheral defoliation; 3: gap-like defoliation; 4: uniform loss of needles; 5: spot-like defoliation; 6: upper crown only; 7: other.

	Defoliation type							
	0	1	2	3	4	5	6	7
Percentage	27	12	0	12	39	2	3	3

Table 10. Percentages of broadleaves in specific crown pattern categories. 0: 0–15% loss of density; 1: no clear pattern; 2: small gaps; 3: gaps in the lateral branch system; 4: large gaps in lateral branch system; 5: predominantly large gaps; 6: whole or part of crown completely defoliated; 7: other. Figures are percentages of trees in each species in each class (i.e. rows sum to 100).

	Crown pattern type							
	0	1	2	3	4	5	6	7
Oak	22	6	37	23	3	2	2	5
Beech	25	10	35	20	5	0	1	4

system has been questioned by some authors (e.g. Thiebaut, 1988; Athari and Kramer, 1989). Another, used in Table 10, was developed by Westman (1989). A comparison between the two is given in Table 11.

There is clearly a considerable amount of disagreement between the two techniques. Over half of the oaks classified as 0 in the Westman scheme were classified as 1 under the Roloff scheme. The reverse is also true, with many of

Table 11. Comparison of crown patterns using the Westman and Roloff techniques. Westman scores: 0: less than 20% loss of density; 1: no clear pattern; 2: small gaps; 3: gaps in the lateral branch system; 4: large gaps; 5: mainly large gaps; 6: whole or part of crown completely defoliated; 7: other. Roloff scores: 0: vigorous growth of both apical and side shoots; 1: side shoots growing more slowly than apical shoots; 2: reduced shoot growth; 3: growth stopped and dieback occurring. The numbers of trees in each category are given.

Oak	Westman score							
	0	1	2	3	4	5	6	7
Roloff score								
0	140	14	68	18	3	0	0	15
1	200	61	362	139	4	1	2	48
2	41	31	215	212	37	20	20	18
3	0	2	3	23	8	22	6	2

Beech	Westman score							
	0	1	2	3	4	5	6	7
Roloff score								
0	126	14	58	3	2	1	0	5
1	84	49	163	86	14	1	7	13
2	4	12	71	67	11	0	3	17
3	0	13	6	16	13	1	1	1

the trees classified as 0 under the Roloff scheme having values of 1 or more in the Westman scheme. The results for beech are similar, although a slightly higher proportion were scored as 0 in both schemes.

Dead shoots in conifers

Shoot death in conifers was assessed on the basis of location within the crown, location on branches and frequency. In all three species, the extent of shoot death increased from 1989 (Table 12). Shoot death was recorded in over half of the trees assessed in 1990. However, there is no evidence that this represents an unusual situation.

As in 1989, the extent of shoot death increased with decreasing levels of crown density. As the relationships in 1990 were very similar to those for 1989 (presented in Innes and Boswell, 1990), tables illustrating them have not been repeated here.

Table 12. Extent of shoot death in the live crowns of conifers, 1990. The percentage of trees in each species in a given category is shown.

	Percentage frequency				
	Absent	Rare	Scattered	Common	Abundant
Sitka spruce	46	8	18	28	2
Norway spruce	47	10	19	23	1
Scots pine	43	22	23	10	1

Crown dieback in broadleaves

The extent of crown dieback provides a good measure of the health of a tree. Consequently, considerable effort has gone into devising suitable ways of assessing dieback. The system used here is based on Westman (1989) and involves assessing the degree (Table 13a), location and extent (Table 13b) of dieback.

There was an increase in the degree of dieback in both beech and oak in 1990. This mainly involved trees classified as being without dieback in 1989, many of which were recorded as having dieback of thin branches in 1990 (Table 13d). There were also changes in the location of dieback (Table 13e). In the case of oak, there was no clear pattern in the changes to dieback location. With beech, the majority of changes were related to an increase of dieback of branches in the tops and the top and middle of trees.

As might be expected from the above, the extent of dieback increased (Tables 13f and 13g). In oak, there was a big increase in the number of trees moving from class 0 (no dieback) to 5% dieback. However, there were also considerable numbers showing a marked worsening, with up to 70% dieback recorded in trees scored with no dieback in 1989. This represents a potentially

Table 13 (a). Percentages of oak and beech in each crown dieback class, 1990. 0: no dieback; 1: leaf loss only; 2: breaks to thin branches; 3: breaks to thick branches; 4: stem broken; 5: other. The percentage of trees in each species in a given category is shown.

	Class					
	0	1	2	3	4	5
Oak	33	16	43	7	0	0
Beech	24	31	41	3	0	0

(b). Location of dieback in oak and beech. 0: none; 1: top of tree only; 2: middle parts of crown; 3: top and middle; 4: base of crown; 5: throughout crown.

	Class						
	0	1	2	3	4	5	6
Oak	33	9	10	15	7	26	0
Beech	24	33	5	20	2	15	1

(c). Extent of dieback (in 10% classes). The percentage of trees of each species in a given category is shown.

	0	1-10	11-20	21-30	31-40	41-50	51-60	61-70	71-80	81-90	91-100
Oak	35	51	10	2	1	1	0	0	0	0	0
Beech	27	57	12	3	1	0	0	0	0	0	0

Table 13d. Changes in the degree of dieback between 1989 and 1990. 0: none; 1: leaf loss only; 2: dieback restricted to relatively thin branches; 3: several large branches involved; 4: main stem involved. Numbers of trees in each category are given.

	Score in 1989					
Score in 1990	0	1	2	3	4	5
Oak						
0	**304**	59	27	2	0	0
1	85	**51**	65	11	1	0
2	158	75	**230**	56	1	0
3	17	4	16	**60**	4	1
4	0	1	1	1	**1**	0
5	0	0	1	0	0	**0**
Beech						
0	**165**	33	8	0	0	0
1	116	**112**	20	1	0	0
2	107	116	**123**	4	0	1
3	6	1	4	**16**	0	0
4	0	0	0	1	**2**	0

Table 13e. Changes in the location of dieback between 1989 and 1990. 0: none, 1: top of tree only; 2: middle parts of crown; 3: top and middle; 4: base of crown; 5: throughout crown; 6: other. Numbers of trees in each category are given.

	Location in 1989						
Location in 1990	0	1	2	3	4	5	6
Oak							
0	**304**	30	21	14	9	14	0
1	54	**23**	3	3	5	6	0
2	52	20	**25**	13	4	17	0
3	54	34	18	**36**	3	26	0
4	44	6	3	10	**12**	9	0
5	56	20	20	40	6	**218**	0
Beech							
0	**165**	20	5	7	8	1	0
1	133	**108**	7	17	2	9	0
2	23	4	**5**	4	2	1	0
3	33	52	10	**43**	4	19	0
5	1	5	1	2	**3**	1	0
5	39	6	2	17	2	**65**	0
6	0	4	0	3	0	3	**0**

Table 13f. Changes in the extent of oak dieback between 1989 and 1990. Numbers of trees in each category are given.

	Dieback extent (%) in 1989																
Dieback extent (%) in 1990	0	5	10	15	20	25	30	35	40	45	50	55	60	65	70	75	80
0	**348**	31	24	7	4	1	0	0	0	0	0	0	0	0	0	0	0
5	198	**80**	126	37	13	2	3	1	0	0	1	0	0	0	0	0	0
10	33	21	**57**	24	12	4	5	1	2	1	0	0	0	0	0	0	0
15	14	3	17	**34**	11	11	4	1	0	0	0	1	0	0	0	0	0
20	6	1	2	4	**11**	8	1	2	4	0	2	0	0	0	0	0	0
25	0	0	3	0	2	**2**	1	1	0	1	0	0	0	0	0	0	1
30	2	0	0	1	2	2	**3**	0	0	0	1	0	1	0	0	0	0
35	1	0	0	0	0	0	0	**2**	1	0	0	0	0	0	0	0	0
40	3	0	0	0	0	0	2	1	**0**	0	1	0	0	0	0	0	1
45	1	1	0	0	0	0	0	0	0	**0**	0	0	0	0	0	0	0
50	9	0	1	0	0	1	0	0	1	0	**0**	0	0	0	0	0	0
55	1	0	1	0	0	0	0	1	0	0	0	**0**	0	0	0	0	0
60	1	0	0	0	0	0	0	0	0	1	0	0	**0**	0	0	0	0
65	0	0	0	1	0	0	1	0	0	0	0	0	0	**0**	0	0	0
70	1	0	0	0	0	0	0	0	0	0	0	0	0	0	**0**	0	0
75	0	0	0	0	0	0	0	0	0	0	0	0	0	0	0	**0**	1
80	0	0	0	1	0	0	0	0	0	0	0	0	0	0	0	0	**0**

Table 13g. Changes in dieback extent of beech between 1989 and 1990. Numbers of trees in each category are given.

Dieback extent (%) in 1990	Dieback extent (%) in 1989										
	0	5	10	15	20	25	30	35	40	45	50
0	**197**	21	10	1	1	0	0	0	0	0	0
5	173	**87**	52	18	4	1	0	0	0	0	0
10	35	26	**34**	27	9	1	2	0	0	1	0
15	12	3	10	**26**	11	2	1	0	0	0	1
20	13	0	2	5	**11**	4	0	1	0	0	0
25	2	0	0	0	3	**1**	1	3	0	0	0
30	9	0	1	0	0	0	**2**	2	0	0	0
35	1	0	0	0	0	0	1	**0**	0	0	0
40	4	0	0	0	1	0	1	0	**0**	0	0
45	1	0	0	0	0	0	0	0	0	**0**	0
50	0	0	0	0	0	0	0	0	0	0	**0**
55	0	0	0	0	0	0	0	0	0	0	0
60	0	0	0	0	0	0	0	0	0	0	0
65	1	0	0	0	0	0	0	0	0	0	0

worrying situation, particularly in view of increased reports of dieback from central and northern Europe (Eichholz, 1985; Balder and Lakenberg, 1987; Donaubauer, 1987; Jakucs, 1988; Hartmann et al., 1989). A number of trees showed recovery over the year, particularly those with only a little dieback in 1989. Observer error may be involved.

The extent of dieback in beech was much less than in oak and this is reflected in Table 13g. Again, there was evidence for quite serious deterioration in a number of trees that were scored as having no dieback in 1989. Trees with 10% or more dieback in 1989 on average showed signs of recovery, suggesting a complex and dynamic pattern of deterioration and recovery.

Discoloration

As already indicated in the section on changes in forest condition during the monitoring programme, foliage discoloration was one of the most marked effects of the drought. The amount of discoloration increased between the beginning of July and the end of August at drought-stressed sites, and the amount recorded at some sites will therefore be partially dependent on the date of assessment. The amounts of discoloration recorded in conifers and broadleaves is presented in Table 14. In the conifers, discoloration was divided according to the age of needles affected, the two categories being current-year needles and older needles.

Given the increase in discoloration recorded by the repeat assessments, the figures presented in Table 14 are rather surprising. In virtually every case (the exceptions being browning of Scots pine needles and yellowing of current-year Sitka spruce needles), discoloration values were lower in 1990 than in 1989 (Tables 14f and 14h).

Sitka spruce

In Sitka spruce, the browning recorded on trees in 1989 was absent in 1990, although a few trees developed a small amount of browning. There was a substantial increase in the amount of yellowing of current-year needles. This mainly involved trees scored as without yellowing in 1989; those that had yellowing in 1989 largely recovered. The increase was restricted to current-year needles; very few trees developed yellowing of older needles and trees with yellow older needles in 1989 showed a marked recovery. The overall discoloration values reflect both the development of yellowing in some trees and the general recovery of those trees affected in 1989.

Table 14. Percentages of trees in each discoloration class.
(a). Percentages of trees in each needle-browning class, 1990.

	Current					Old				
	0-10%	11-25%	26-60%	>60%	Dead	0-10%	11-25%	26-60%	>60%	Dead
Sitka spruce	99	1	0	0	0	99	1	0	0	0
Norway spruce	98	1	0	0	0	98	1	1	0	0
Scots pine	91	8	0	0	1	91	7	1	0	1

(b). Percentages of trees in each needle-yellowing class, 1990.

	0-10%	11-25%	26-60%	>60%	Dead	0-10%	11-25%	26-60%	>60%	Dead
Sitka spruce	94	3	2	1	0	98	1	0	0	0
Norway spruce	98	1	0	0	0	99	0	0	0	0
Scots pine	97	3	0	0	1	95	4	0	0	1

(c). Percentages of trees in each overall discoloration class, 1990.

	Overall discoloration				
	0-10%	11-25%	26-60%	61-99%	Dead
Sitka spruce	94	4	2	0	0
Norway spruce	97	3	0	0	0
Scots pine	88	10	1	0	1

(d). Percentages of oak and beech showing leaf discoloration, 1990.

	Browning					Yellowing				
	0-10%	11-25%	26-60%	>60%	Dead	0-10%	11-25%	26-60%	>60%	Dead
Oak	95	4	0	0	0	92	6	1	1	0
Beech	91	5	4	0	0	88	8	3	1	0

(e). Percentages of beech and oak showing overall discoloration, 1990.

	Class				
	0-10%	11-25%	26-60%	>60%	Dead
Oak	90	8	2	1	0
Beech	86	9	3	2	0

Table 14f. Changes in discoloration of Sitka spruce between 1989 and 1990. The numbers of trees in each category are given.

	1989 Browning scores								
	Current needles					Older needles			
1990 scores	0-10%	11-25%	26-60%	>60%		0-10%	11-25%	26-60%	>60%
0-10%	**1288**	21	4	2		**1272**	23	11	6
11-25%	9	**0**	0	0		8	**0**	1	1
26-60%	0	0	**0**	0		0	0	**0**	0
>60%	0	0	0	**0**		1	0	0	**0**

	Current needles				1989 Yellowing scores	Older needles			
1990 scores	0-10%	11-25%	26-60%	>60%		0-10%	11-25%	26-60%	>60%
0-10%	**1222**	8	5	1		**1222**	47	21	4
11-25%	31	**5**	2	0		10	**4**	6	0
26-60%	29	1	**2**	0		3	3	**0**	0
>60%	10	0	0	**1**		2	0	0	**0**

	1989 Overall discoloration			
1990 scores	0-10%	11-25%	26-60%	>60%
0-10%	**1136**	75	22	6
11-25%	38	**12**	8	2
26-60%	13	4	**4**	2
>60%	2	0	0	**0**

Norway spruce

Very little discoloration of current needles was recorded on Norway spruce in either 1989 or 1990. Trees with browning of older needles in 1989 generally recovered, although a small proportion of trees developed browning. The yellowing of older needles seen in 1989 was absent in 1990. The changes are reflected in the overall discoloration scores, with a few trees developing discoloration and most of those with discoloration in 1989 recovering.

Scots pine

Discoloration was more frequent on Scots pine than on either of the spruce species. This is thought to be normal, reflecting the onset of autumnal senescence, which is much more marked in Scots pine than in the spruces. Although there was an increase in the amount of browning recorded, this took the form of an increase in the number of trees affected rather than an increase in degree. As with the spruces,

Table 14g. Changes in discoloration of Norway spruce between 1989 and 1990. Numbers of trees in each category are given.

	Current needles				1989 Browning scores	Older needles			
1990 scores	0-10%	11-25%	26-60%	>60%		0-10%	11-25%	26-60%	>60%
0-10%	**1672**	5	2	0		**1590**	63	19	1
11-25%	21	**2**	0	0		15	**1**	3	0
26-60%	2	0	**0**	0		9	2	**0**	0
>60%	0	1	0	**0**		2	0	0	**0**

					1989 Yellowing scores				
1990 scores	0-10%	11-25%	26-60%	>60%		0-10%	11-25%	26-60%	>60%
0-10%	**1654**	17	8	2		**1656**	30	8	0
11-25%	14	**1**	0	0		6	**1**	1	0
26-60%	5	0	**0**	0		0	0	**0**	0
>60%	4	0	0	**0**		3	0	0	**0**

	1989 Overall discoloration scores			
1990 scores	0-10%	11-25%	26-60%	>60%
0-10%	**1544**	78	27	3
11-25%	36	**3**	5	0
26-60%	4	1	**0**	0
>60%	3	1	0	**0**

Table 14h. Changes in discoloration scores of Scots pine between 1989 and 1990. Numbers of trees in each category are given.

1990 scores	1989 Browning scores							
	Current needles				Older needles			
	0-10%	11-25%	26-60%	>60%	0-10%	11-25%	26-60%	>60%
0-10%	**1557**	98	3	1	**1589**	88	7	0
11-25%	124	**21**	0	0	74	**45**	10	0
26-60%	1	1	**1**	0	11	6	**2**	0
>60%	1	0	0	**0**	0	0	0	**0**

1990 scores	1989 Yellowing scores							
0-10%	**1726**	55	0	1	**1582**	129	36	0
11-25%	49	**2**	1	0	56	**18**	7	0
26-60%	0	0	**0**	0	3	3	**1**	0
>60%	0	0	0	**0**	1	1.	0	**0**

1990 scores	1989 Overall discoloration scores			
0-10%	**1418**	179	33	0
11-25%	123	**59**	12	0
26-60%	6	6	**3**	0
>60%	0	0	0	**0**

trees with discoloration in 1989 mostly recovered and the observed discoloration in 1990 was on trees previously classified as being without any.

Oak

Oak showed relatively little change over the period. A similar pattern to the conifers occurred, with discoloration developing on trees that had previously been unaffected and affected trees recovering. A small number of trees recorded as having limited (11–25% of the foliage) yellowing in 1989 deteriorated further.

Beech

The greatest changes occurred in beech, with marked reductions in the amount of browning and yellowing. There was some evidence for the new development of browning in trees, although the proportion involved was relatively small, particularly in comparison to the number of trees recovering.

Needle retention

Needle retention provides an index of the number of years that needles are retained for. It only provides an indication of needle loss if the loss occurs progressively from the oldest needles to the youngest needles. As it is difficult to count back more than about 7 years, an upper limit of >7 years is used. Above 7 years, there may be substantial variation between trees but the assessment errors are proportionally greater. The index is subject to error as needle retention varies throughout the crown and observers have to make an estimate of the average length of retention. Consequently, increases of more than one year retention (which are obviously impossible) may be recorded. Although quite detailed indices of needle retention have been proposed (e.g. Becher, 1986), the errors associated with the estimates of retention mean that these are of relatively little significance. Consequently, average values have been used. The results for 1990 are shown in Table 15a.

Considerable changes in the pattern of

Table 14i. Changes in discoloration of oak and beech between 1989 and 1990. Numbers of trees in each category are given.

Oak 1990 score	Browning in 1989				Yellowing in 1989			
	0-10%	11-25%	26-60%	>60%	0-10%	11-25%	26-60%	>60%
0-10%	**1122**	39	0	4	**1048**	52	16	8
11-25%	49	**10**	1	0	55	**8**	2	2
26-60%	5	0	**0**	0	18	5	**0**	0
>60%	0	0	0	**0**	11	2	4	**0**

Beech 1990 score								
0-10%	**572**	153	27	3	**539**	148	43	5
11-25%	22	**15**	5	2	36	**22**	6	3
26-60%	25	3	**4**	0	8	6	**6**	2
>60%	4	0	0	**0**	6	4	0	**1**

	Overall discoloration							
	Oak 1989				Beech 1989			
1990 score	0-10%	11-25%	26-60%	>60%	0-10%	11-25%	26-60%	>60%
0-10%	**1011**	57	12	7	**478**	162	65	15
11-25%	83	**14**	5	2	37	**18**	15	2
26-60%	21	4	**1**	0	11	8	**8**	1
>60%	9	5	1	**0**	6	6	2	**1**

Table 15a. Percentages of trees with needle life (needle retention) for a given number of years, 1990.

	Number of years								
	0	1	2	3	4	5	6	7	>7
Sitka spruce	0	1	2	4	7	11	19	24	31
Norway spruce	0	0	0	2	6	13	21	24	32
Scots pine	1	3	50	45	1	0	0	0	0

needle retention occurred between 1989 and 1990 (Tables 15b and 15c).

Many Sitka spruce deteriorated during the year, with the changes being most apparent in the trees with higher needle retention. For example, 212 (37%) of the trees recorded as having more than 7 years' needle retention in 1989 had 7 years or less in 1990. Similarly, whereas 48 (20%) of the trees recorded with 7 years' needle retention in 1989 had more in 1990, 79 (33%) had less.

The changes were even more marked in Norway spruce, which was not affected to the same extent by the 1989 aphid defoliation episode and therefore had fewer severely defoliated trees available for recovery. Trees with only 2 to 4 years' needle retention in 1989 generally recovered whereas substantial numbers of trees with relatively high levels of needle retention deteriorated. A number of trees with 6 or more years of foliage in 1989 had 3 or less years in 1990.

Table 15b. Changes in needle retention of Sitka and Norway spruce between 1989 and 1990.

Sitka spruce	Needle retention in 1989 (years)							
1990 retention	1	2	3	4	5	6	7	>7
1	**7**	0	0	1	0	0	0	0
2	20	**6**	4	1	0	2	0	0
3	1	14	**14**	15	4	1	3	0
4	0	5	7	**31**	19	8	4	3
5	0	0	5	21	**61**	29	20	14
6	0	0	3	5	54	**70**	52	63
7	0	0	0	2	10	60	**111**	132
>7	0	0	0	0	2	15	48	**354**

Norway spruce								
1	**3**	0	2	0	0	1	0	1
2	0	**0**	0	3	0	0	1	0
3	0	1	**10**	9	3	6	2	9
4	0	0	7	**37**	33	14	7	10
5	0	0	3	19	**97**	50	26	29
6	0	0	0	1	44	**113**	97	101
7	0	0	1	0	2	61	**165**	175
>7	0	0	0	0	1	6	46	**507**

Table 15c. Changes in needle retention of Scots pine between 1989 and 1990.

	Needle retention in 1989 (years)			
1990 retention	1	2	3	4
1	**19**	22	5	0
2	20	**684**	213	4
3	0	236	**589**	21
4	0	1	18	**6**

Scots pine was relatively stable overall, with almost equal number of trees with 2 years' needle retention increasing to 3 years and vice-versa.

Leader condition

Leader condition was assessed in conifers and the results are presented in Table 16. There were no marked changes from 1989, although the number of double leaders on Scots pine was lower in 1990.

Flowering in Scots pine

Male flowering replaces needle development and can result in a thinner-looking crown. The gaps on the shoots left after the flowers have dropped off were assessed on each tree, with trees being divided into upper and lower crowns (Table 17a). As in 1989, flowering was much

Table 16. Percentages of trees with the leader in a given condition, 1990. 0: normal; 1: shorter than current-year side shoots; 2: missing; 3: bent or twisted; 4: double; 5: broken; 6: bare; 7: side shoots taken over; 8: complete loss of apical dominance.

	Type								
	0	1	2	3	4	5	6	7	8
Sitka spruce	65	4	9	10	4	0	0	6	0
Norway spruce	65	6	16	4	4	1	1	5	0
Scots pine	62	4	5	1	5	0	1	11	10

more common in the lower crown than in the upper part (Table 17b). The figures for 1989 and 1990 were similar, although the amount of flowering in the upper crown was less in 1990.

As with some of the other indices, there was considerable variation between trees (Table 17c). The extent of flowering in one year is not a good indication of the extent in the following year.

To a certain extent, the amount of flowering can be related to crown density (Table 17d). Heavy flowering in both the top and bottom of the crown was associated with lower crown densities. It is uncertain whether the heavier flowering occurs as a response to the low crown densities or whether the thinness associated with the flowering is the reason why the trees were scored as having lower densities. There was no indication that flowering was more common on trees with very thin crowns, suggesting that the latter explanation may be more likely, but this requires further investigation.

Another potential index of stress is the defoliation type. If flowering is a sign of stress, then it should be more apparent in trees showing high (i.e. poor) values for defoliation type. However,

Table 17a. Numbers of Scots pine with given scores for flowering in the top and base of the crown, 1990.

	Flowering score for top half				
Flowering in bottom half	0	1	2	3	4
0	**226**	6	1	1	0
1	338	**35**	5	0	1
2	341	92	**16**	0	0
3	247	182	66	**10**	0
4	73	97	135	66	**6**

Table 17b. Percentage of pine in each flowering class, 1990. 0: none; 1: rare; 2: infrequent; 3: common; 4: abundant.

	Flowering class				
	0	1	2	3	4
Upper crown	63	21	11	4	0
Lower crown	12	19	23	26	19

Table 17c. Changes in the extent of flowering on individual Scots pine between 1989 and 1990. The classes are the same as in Table 17b.

	Flowering class in upper crown in 1989				
Flowering in upper crown 1990	0	1	2	3	4
0	**721**	281	103	35	5
1	163	**129**	71	36	4
2	53	47	**58**	53	7
3	16	12	20	**22**	6
4	3	1	0	1	**2**
	Flowering class in lower crown in 1989				
Flowering in lower crown 1990	0	1	2	3	4
0	**79**	62	45	20	9
1	69	**108**	88	58	26
2	62	89	**132**	101	41
3	32	63	126	**154**	118
4	10	31	41	94	**192**

the data presented in Table 17d do not reveal any evidence of a trend for increased flowering amongst the higher values.

Fruiting

The amount of fruiting (coning, acorns and mast) recorded in 1990 is given in Table 18a. In the conifers, only fresh cones were assessed. In all cases, the figures refer to the amount of visible fruit, and no account has been taken of the quality of the seed.

Coning was more frequent in the two spruce species in 1990 than in 1989. The increase involved both trees without cones in 1989 and trees with rare or scattered cones. Coning in Scots pine was generally less, although increases were recorded on a substantial proportion of the trees. The amount of acorns on oak was much less than in 1989 although, again, increases occurred on some trees. The reduction can be attributed to the loss of flowers due to spring frosts in April 1990. The amount of beech mast was particularly apparent in 1990. It was much more frequent than in 1989 (Table 18b), with many trees having mast that were without

33

Table 17d. Relationship between flowering and crown density in Scots pine, 1990. Figures are the percentage of trees in each crown density category with a given level of flowering (i.e. columns sum to 100%).

Crown density score	Flowering – top of crown					Flowering – base of crown				
	0	1	2	3	4	0	1	2	3	4
0	2.5	1.5	0.9	1.3	0	2.6	3.4	2.2	1.2	1.3
1	4.2	2.2	2.2	3.9	0	7.7	4.7	2.9	2.8	1.3
2	8.4	6.6	4.9	3.9	0	9.8	11.6	7.8	5.1	4.2
3	12.0	13.6	8.5	10.4	0	12.4	11.9	13.6	11.3	10.1
4	17.5	18.0	13.9	20.8	14.3	15.8	21.1	17.8	14.5	17.5
5	16.2	20.1	19.3	13.0	0	15.8	16.6	14.9	19.2	18.8
6	13.7	15.8	20.2	10.4	14.3	6.0	11.9	14.7	19.4	17.0
7	9.0	12.4	12.6	14.3	28.6	5.6	8.7	11.6	10.7	13.3
8	6.3	5.1	9.4	10.4	14.3	4.7	4.7	6.7	6.7	9.3
9	2.8	2.7	4.0	2.6	0	1.7	0.8	3.6	4.4	2.9
10	2.0	0.5	0.9	2.6	0	2.1	1.3	1.8	1.6	1.3
11	0.7	0.5	0.9	3.9	0	1.3	0.8	0.2	0.8	1.3
12	1.1	0.2	0.9	2.6	14.3	0.4	1.3	0.9	1.0	1.1
13	0.3	0	0	0	0	0.4	0.5	0	0.2	0
14	0.3	0.2	0	0	0	0.4	0.3	0.2	0.4	0
15	0.5	0	0	0	0	1.3	0.3	0.4	0	0
16	0.7	0.5	0.9	0	14.3	2.6	0	0.7	0.4	0.5
17	0.2	0	0	0	0	1.3	0	0	0	0
18	0.1	0	0.4	0	0	0.4	0	0	0.2	0
19	0.4	0.2	0	0	0	2.1	0	0	0.2	0
20	1.1	0	0	0	0	5.6	0	0	0	0

Table 17e. Relationship between defoliation type and flowering in Scots pine. Figures are the percentage of trees with a given level of flowering in a given defoliation category (i.e. rows sum to 100%).

Defoliation type	Flowering – top of crown					Flowering – base of crown				
	0	1	2	3	4	0	1	2	3	4
0	69.3	19.6	7.9	3.2	0	17.1	24.9	23.5	21.1	13.4
1	63.4	18.1	14.5	3.1	0.9	11.0	11.5	22.9	31.3	23.3
2	16.7	33.3	50.0	0	0	0	16.7	0	50.0	33.3
3	58.4	28.6	9.4	3.7	0	9.0	24.9	22.4	28.2	15.5
4	60.9	21.4	12.9	4.6	0.3	9.3	16.9	23.2	27.2	23.3
5	50.0	29.4	14.7	5.9	0	5.9	5.9	23.5	41.2	23.5
6	62.1	12.1	18.2	4.5	3.0	13.6	16.7	24.2	24.2	21.2
7	64.7	19.1	8.8	5.9	1.5	20.6	23.5	22.1	16.2	17.6

it in 1989. In addition, trees with light levels of mast in 1989 had higher levels in 1990.

It is generally held that fruiting occurs to a certain extent in response to stress. The general increase in fruiting in 1990 suggests that some environmental factor was operative, but it is not clear which of a variety of factors it was. Heavy fruiting of some deciduous species (such as oak and sweet chestnut) was recorded in 1989, suggesting that drought conditions during the previous summer was not the only factor affecting fruiting. This is borne out by the detailed research undertaken in Scandinavia, which has not determined any clear relationships between specific environmental variables and the incidence of fruiting (e.g. Hagner, 1956).

Table 18a. Percentages of trees with a given level of fruiting. For spruce, only fresh cones were included. For Scots pine, only unopened second year cones were included.

	Frequency			
	None	Scarce	Common	Abundant
Sitka spruce	36	34	25	6
Norway spruce	51	25	20	5
Scots pine	24	36	32	8
Oak	74	16	6	3
Beech	6	11	33	50

very low levels of coning.

In oak, the low frequency of acorns in 1990 meant that few conclusions could be reached about the relationship between crown density and acorn production. Beech showed a very strong relationship between mast production and crown density. Only five trees with 35% or more loss of density were without mast and virtually all of the severely defoliated trees had high levels of mast. This suggests that in beech there may be a relationship between the amount of stress and the amount of mast.

Table 18b. Changes in the amount of fruiting between 1989 and 1990.

Sitka spruce 1990 score	Score in 1989					Norway spruce 1990 score	Score in 1989			
	0	1	2	3			0	1	2	3
0	**441**	29	8	0		0	**834**	14	8	2
1	294	**90**	44	13		1	376	**53**	5	0
2	148	115	**50**	17		2	277	44	**13**	0
3	32	22	13	**8**		3	51	22	6	**0**

Scots pine 1990 score	0	1	2	3		Oak 1990 score	0	1	2	3
0	**173**	172	74	19		0	**562**	209	114	42
1	97	**280**	216	88		1	83	**65**	24	17
2	37	208	**217**	123		2	35	23	**17**	7
3	5	21	49	**69**		3	12	10	9	**3**

Beech 1990 score	0	1	2	3
0	**40**	6	0	1
1	67	**26**	1	0
2	167	77	**27**	10
3	218	130	48	**18**

The relationship between crown density and the degree of fruiting is shown in Tables 18c and 18d. In the conifers, there was little evidence that increased coning was associated with thinner crowns, although particularly dense Sitka spruce crowns had a greater frequency of coning than less dense ones. In all three species, severely defoliated crowns were associated with

Secondary shoots in spruce

Secondary shoots were assessed in both 1989 and 1990. However, during the course of the quality assurance work in 1989, it became apparent that there were problems with variation in the ways that secondary shoots were being assessed. Consequently, no figures were given in the 1989 report.

Table 18c. Relationship between crown density and coning in the three conifers, 1990. The figures refer to the percentages of trees in a given crown density class with a given amount of coning (i.e. rows sum to 100).

Crown density score	Sitka spruce				Coning Norway spruce				Scots pine			
	0	1	2	3	0	1	2	3	0	1	2	3
0	34.6	7.7	26.9	30.8	51.5	20.8	21.8	5.9	25.0	20.0	20.0	35.0
1	39.5	18.6	18.6	23.3	58.5	18.6	16.1	6.8	17.6	32.4	36.8	13.2
2	40.0	32.5	25.0	2.5	54.8	23.8	15.5	6.0	15.3	38.9	31.9	13.9
3	33.5	33.5	26.4	6.6	58.4	21.4	16.5	3.7	13.9	35.7	37.8	12.6
4	33.9	35.6	26.4	4.0	54.6	22.9	19.7	2.8	19.3	35.4	38.1	7.1
5	33.0	29.2	31.9	5.9	49.6	21.2	22.9	6.4	22.4	34.6	34.6	8.4
6	34.4	35.6	26.4	3.7	45.9	28.2	21.1	4.8	21.0	37.4	36.4	5.2
7	28.1	38.4	30.1	3.4	49.4	29.7	16.9	4.1	27.9	44.8	25.9	1.5
8	40.2	40.2	16.8	2.8	38.3	29.8	27.7	4.3	21.9	50.0	23.4	4.7
9	33.7	34.9	24.1	7.2	28.3	41.5	28.3	1.9	39.3	35.7	21.4	3.6
10	42.9	40.5	11.9	4.8	37.5	35.0	22.5	5.0	45.2	35.5	19.4	0
11	50.0	32.1	10.7	7.1	45.0	40.0	15.0	0	68.8	25.0	6.3	0
12	39.3	39.3	21.4	0	44.4	22.2	33.3	0	73.7	21.1	5.3	0
13	40.0	36.0	20.0	4.0	33.3	33.3	27.8	5.6	75.0	25.0	0	0
14	60.0	20.0	20.0	0	71.4	14.3	0	14.3	80.0	20.0	0	0
15	62.5	25.0	12.5	0	0	100.0	0	0	83.3	16.7	0	0
16	83.3	16.7	0	0	33.3	33.3	33.3	0	100.0	0	0	0
17	50.0	50.0	0	0	33.3	66.7	0	0	66.7	0	33.3	0
18	100.0	0	0	0	100.0	0	0	0	100.0	0	0	0
19	100.0	0	0	0	100.0	0	0	0	100.0	0	0	0
20	100.0	0	0	0	100.0	0	0	0	100.0	0	0	0

Secondary shoots were assessed on the basis of abundance and location, following the recommendations of (Lesinski, 1989). The relative frequency of location is given in Table 19a and their abundance is given in Table 19b.

The patterns for the two species are remarkably similar, despite the differences in defoliation levels. The presence of secondary shoots is believed to be a good index of vigour, indicating that the tree is capable of recovery from a defoliating event (Gruber, 1987; Lesinski and Westman, 1987; Lesinski and Landmann, 1988; Rehfuess and Rehfuess, 1988; Lesinski, 1989; Rehfuess, 1989). Consequently, the relationship between crown density and the occurrence of secondary shoots is of interest, and has been presented in Table 19c.

Again, the two sets of results are fairly similar, although the proportion of Norway spruce in the lower defoliation categories without secondary shoots is lower than for Sitka spruce. As much of the defoliation is known to have occurred fairly recently (1988–89), it is possible that secondary shoots have yet to develop on many of the trees. This should become apparent in future years.

Epicormic branches on oak

To a certain extent, epicormic shoots on oak serve a similar function to secondary shoots on spruce. The proportions of trees with branch and stem epicormics is given in Table 20.

Epicormics were generally more common on the branches than on stems. This may reflect the effect of pruning or a real trend. Trees without branch epicormics never had stem epicormics.

Leaf-rolling in beech

Leaf-rolling is believed to be associated with water stress and this appears to be confirmed

Table 18d. Relationship between crown density and the degree of fruiting in oak and beech, 1990. Figures refer to the percentages of trees in a given crown density class with a given level of fruiting (i.e. rows sum to 100).

	\multicolumn{8}{c}{Degree of fruiting}							
	Oak				Beech			
Crown density score	0	1	2	3	0	1	2	3
0	51.7	24.1	13.8	10.3	33.3	16.7	33.3	16.7
1	50.7	26.1	13.0	10.1	23.1	17.9	25.6	33.3
2	60.8	17.7	14.6	7.0	12.2	20.3	32.4	35.1
3	62.7	16.3	12.7	8.4	4.1	6.8	33.8	55.4
4	64.7	26.9	6.5	2.0	4.8	9.5	35.2	50.5
5	75.9	15.6	5.5	3.0	2.6	12.2	40.0	45.2
6	76.2	17.7	5.2	0.9	3.6	8.0	34.8	53.6
7	83.0	11.6	3.5	1.9	0.9	9.4	28.3	61.3
8	80.3	15.6	4.0	0	1.5	9.0	32.8	56.7
9	87.2	11.7	0	1.1	0	18.5	38.9	42.6
10	86.9	9.8	3.3	0	0	15.4	33.3	51.3
11	84.2	15.8	0	0	8.0	4.0	20.0	68.0
12	96.3	3.7	0	0	0	20.0	60.0	20.0
13	95.2	0	4.8	0	0	0	11.1	88.9
14	76.5	17.6	5.9	0	0	0	0	100.0
15	70.0	0	20.0	10.0	0	0	0	100.0
16	100.0	0	0	0	0	0	0	100.0
17	100.0	0	0	0	33.3	0	33.3	33.3
18	0	0	0	0	0	0	0	100.0
19	100.0	0	0	0	0	0	0	0

Table 19a. Location of secondary shoots on Sitka and Norway spruce in 1990. 0: none; 1: outer parts of branches only; 2: middle of branches; 3: inner parts of branches; 4: outer and middle parts; 5: middle and inner parts; 6: all along branch. Figures are percentages.

	\multicolumn{7}{c}{Location of secondary shoots}						
	0	1	2	3	4	5	6
Sitka spruce	54	5	19	0	11	6	5
Norway spruce	56	6	19	0	13	2	3

Table 19b. Abundance of secondary shoots on Sitka and Norway spruce in 1990. 0: none; 1: a few shoots on less than a quarter of the branches; 2: many shoots on less than a quarter of branches; 3: a few shoots on many branches; 4: many shoots on many branches.

	\multicolumn{5}{c}{Abundance of secondary shoots}				
	0	1	2	3	4
Sitka spruce	54	14	4	21	7
Norway spruce	56	13	3	21	7

Table 19c. Relationship between crown density and the presence of secondary shoots. The proportion of trees without secondary shoots in each 10% crown density category is given. The occurrence of secondary shoots was not assessed on trees with 20% or less loss of density.

	Crown density									
	0-10	11-20	21-30	31-40	41-50	51-60	61-70	71-80	81-90	>91
Sitka spruce	–	–	40	41	32	32	40	45	100	–
Norway spruce	–	–	29	24	30	34	44	20	43	100

Table 20. Percentages of oak with branch and stem epicormics. 0: none; 1: rare; 2: scattered; 3: common; 4: abundant.

	Frequency category				
	0	1	2	3	4
Stem	16	34	32	15	4
Branch	1	6	31	47	16

by both the repeat assessments and the overall figures for 1990 (Table 21a), which revealed high levels of rolling.

The 1990 assessments indicated an increase in the incidence of leaf-rolling since 1989 but, as with other indices, individual trees showed considerable variation (Tables 21b and 21c). Almost half of the trees scored as without rolling in 1989 showed rolling in 1990, although rolling was also absent on many (34%) trees that were

Table 21a. Percentages of beech with leaf-rolling in the crown, 1990. Extent classes: 0: none; 1: leaves on a few shoots in the upper crown; 2: leaves on a few shoots elsewhere; 3: half of upper crown affected; 4: half of shoots throughout crown affected; 5: most shoots in the upper crown affected; 6: most shoots throughout crown affected; 7: virtually all leaves affected.

	Extent							
Degree	0	1	2	3	4	5	6	7
None	45	0	0	0	0	0	0	0
Slight	0	7	2	12	4	7	2	1
Medium	0	4	0	4	1	5	3	1
Severe	0	0	0	0	0	1	0	0

Table 21b. Changes in the frequency of leaf-rolling in beech between 1989 and 1990. The classes are the same as in Table 21a.

	Leaf-rolling frequency 1989							
Leaf-rolling frequency 1990	0	1	2	3	4	5	6	7
0	**234**	51	4	44	8	24	14	0
1	43	**23**	2	13	1	4	1	0
2	5	1	**4**	2	1	2	2	0
3	61	25	2	**29**	3	7	3	1
4	10	3	0	7	**9**	8	7	0
5	40	16	2	15	5	**22**	8	0
6	11	2	0	6	2	10	**18**	2
7	4	2	0	5	0	5	2	**1**

affected in 1989. The degree of leaf-rolling did not change substantially between the two years, although there were considerable changes amongst individual trees.

Leaf size in beech

Leaf size in beech was recorded (small – normal – large) in 1989 but was not reported as so few trees had other than normal-sized leaves. A particular feature of 1990 was the large number of very small leaves on some trees, particularly in the upper crown. A substantial increase in the number of small-leaved trees was evident from the results (Table 22a).

In the field, there appeared to be an association between the amount of mast and the presence of small leaves. This is confirmed by the data, with 76% of the trees with small leaves having abundant mast. Only 41% of the trees with normal-sized leaves had abundant mast. There was also evidence that the small leaves were associated thin-crowned trees (Table 22b).

Premature leaf loss in beech

Premature leaf loss, indicated by the presence of green leaves under trees was recorded. The values recorded in 1990 were very similar to those in 1989, being infrequent under 17% of trees and common under 9%.

Presence of insects and fungal pathogens

The method of reporting insect and fungal damage on trees was changed in 1990 following the problems experienced in 1989. In 1990, assessments were standardised between species, with five categories being recognised: none, rare, infrequent, common and abundant. In each case, the category refers to the proportion of foliage affected. Consequently, only limited comparisons can be made with scores from 1989.

Sitka spruce and Norway spruce both had similar levels of insect damage, whereas in Scots pine it was much less frequent. Oak had considerably more damage than beech.

No comparison between 1989 and 1990 is possible for the two species. Insect damage to Scots pine was similar in the two years whereas damage to oak increased and damage to beech decreased. A number of factors affect these assessments (such as the extent of mid-season flushing in oak) and any interpretation should be made with care.

Table 21c. Changes in the degree of leaf-rolling in beech between 1989 and 1990.

Degree in 1990	Degree of leaf-rolling 1989			
	None	Slight	Medium	Severe
None	**233**	118	24	3
Slight	126	**131**	27	3
Medium	49	72	**36**	1
Severe	1	11	1	**0**

Table 22a. Comparison of leaf size in beech between 1989 and 1990.

Leaf size in 1990	Leaf size in 1989		
	Normal	Small	Large
Normal	**626**	8	1
Small	179	**19**	0
Large	3	3	**0**

Table 22b. Relationship between the occurrence of small leaves and crown density in beech, 1990.

Leaf size	Crown density (10% classes)									
	0-10	11-20	21-30	31-40	41-50	51-60	61-70	71-80	81-90	>91
Normal	21	23	25	17	9	3	1	1	0	0
Small	2	15	31	27	17	6	1	0	1	0

Table 23. Percentages of trees with insect damage to the foliage, 1990. The frequency scores refer to the percentage of the foliage affected.

	Frequency None	Rare	Infrequent	Common	Abundant
Sitka spruce	56	14	14	12	5
Norway spruce	62	12	13	11	2
Scots pine	70	16	11	3	0
Oak	20	31	31	16	2
Beech	42	23	16	18	2

Fungal damage was also recorded. Only visible damage to foliage was reported, fruiting bodies on the stems or butts of trees being reported under damage to stems and butts. As with insects, reports only assess visible evidence of recent damage at the time of the assessment and it is likely that the observations severely underestimate the true frequency of occurrence. However, although collection of foliar and root samples has been possible in some monitoring programmes (e.g. Anon., 1989a), it has not yet been feasible in the British programme.

Mechanical damage to the crown

Mechanical damage to the crown by wind represents an important cause of crown thinning in Britain. It is present in most years, but was particularly so in 1990, with the effects of storms in October 1987 and January 1990 still very much in evidence. The proportions of trees in different damage categories are given in Table 25.

Damage was most apparent in oak, being recorded on 39% of the trees and least apparent in Norway spruce, with only 17% affected. Abrasion of peripheral shoots (caused by wind) and broken branches caused by wind and/or snow were by far the most common types of damage.

The mechanical damage classification was modified in 1990 so comparison with 1989 is difficult. However, there was a clear increase in the amount of wind damage to conifers. Oak showed little change and there was an increase

Table 24. Frequency of fungal attack, 1990.

	None	Rare	Infrequent	Common	Abundant
Sitka spruce	100	0	0	0	0
Norway spruce	97	1	1	1	0
Scots pine	98	2	0	0	0
Oak	96	1	2	1	0
Beech	97	3	0	0	0

Table 25. Percentages of trees with specific types of mechanical damage within the crown. 0: none; 1: abrasion of peripheral shoots; 2: wind and snow damage; 3: prevailing wind effect; 4: damage caused by adjacent trees leaning into target tree; 5: hail; 6: lightning; 7: damage caused by harvesting operations; 8: other.

	Type 0	1	2	3	4	5	6	7	8
Sitka spruce	79	17	3	0	1	0	0	0	0
Norway spruce	83	14	2	0	0	0	0	0	0
Scots pine	69	20	9	1	0	0	0	0	0
Oak	61	24	13	1	1	0	0	0	0
Beech	74	20	5	0	1	0	0	0	0

in the number of beech with damage. Damage in the form of broken branches was less evident in oak and beech in 1990, suggesting that trees that were not blown over in 1990 suffered relatively little damage and those damaged in the 1987 storm continued to recover.

Butt and stem damage

The scoring for butt and stem damage was also revised in 1990 to take into account experience gained in the 1989 programme, when the assessments were first introduced. Assessments are now the same for all species, but this involved some reductions in the amount of information specific to particular species (e.g. past *Cryptococcus* infestation on beech). Results are presented in Table 26.

The most frequent type of stem damage on Scots pine was caused by squirrels (5% of trees) and involved bark stripping. Tarry patches on the stems were also recorded (3% of trees); many of these would be due to old squirrel damage. Sap flow from cracks or holes was recorded on 11% of beech stems and insect activity on 5%. Sitka and Norway spruce and oak had no particularly frequent forms of stem damage.

Butt damage was most frequent on beech, with 56% of trees affected. Damage by grey squirrels and rabbits was particularly frequent, affecting 12% of trees. Areas of exposed wood, caused by unknown factors, were present on 8% of trees. All species showed relatively high levels of extraction damage to the butts, with Norway spruce being the most affected (9% of trees).

The amount of stem damage recorded in 1990 was very similar to that recorded in 1989, although beech showed an increase. Because of the change in scoring, it is not possible to tell which type of damage increased. Butt damage levels also remained similar, although rather less was recorded in beech in 1990 than in 1989. Given the drought conditions over much of the country and the January storm, an increase in the number of cracks might have been expected. A 1% increase in the percentage of oak with cracks on the stem was recorded, but the incidence of cracks on the stems of the other species either remained constant or declined. The incidence of fresh cracks on the butts either remained constant or decreased in 1990.

Recognition of anomalous or unusual records

The recognition of anomalous or unusual records represents an important aspect of any monitoring programme. There are several dif-

Table 26. Percentages of trees with specific types of butt (below 1.3 m) and stem (above 1.3 m) damage, 1990. 0: none; 1: dead area with bark still present; 2: area of exposed wood; 3: sap/resin flow from cracks or holes; 4: swelling; 5: crack at least 5× long as wide; 6: fungus; 7: extraction damage; 8: insect activity; 9: tarry spots on bark; 10: vandalism; 11: pruning or brashing wounds; 12: animal damage; 13: fire; 14: other.

	Type 0	1	2	3	4	5	6	7	8	9	10	11	12	13	14
Stem															
Sitka spruce	97	0	0	2	0	0	0	0	0	0	0	0	0	0	0
Norway spruce	94	0	0	2	0	1	0	2	0	0	0	1	0	0	0
Scots pine	89	0	1	0	1	0	0	1	0	3	0	0	5	0	0
Oak	96	0	1	0	0	2	0	0	0	0	0	1	0	0	0
Beech	72	1	3	11	2	2	0	0	5	0	1	1	1	2	0
Butt															
Sitka spruce	89	0	0	2	0	0	0	5	0	0	1	2	0	0	0
Norway spruce	81	0	1	3	0	1	0	9	0	0	2	1	2	0	0
Scots pine	94	0	1	0	0	0	0	4	0	0	0	0	0	0	0
Oak	90	1	2	0	1	2	0	3	0	0	1	0	0	0	1
Beech	56	2	8	1	2	5	1	3	5	1	3	1	12	0	0

ferent approaches that can be adopted. Within any given year, trees with one or more anomalous records can be identified. Alternatively, sites where the average values for one or more indices are anomalous can be determined. Both of these approaches are relatively straightforward and can be addressed using the central limits theorem, as was done in 1989 (Innes and Boswell, 1990).

With the collection of several years' data, a more sophisticated approach is possible. Again, either individual tree records or mean site values can be used. However, a major problem exists with the establishment of a base year which can then be used to draw up comparisons with other years. Some of these problems have already been addressed by epidemiologists (e.g. Raubertas, 1988; Raubertas et al., 1989; Shaw et al., 1988) and others (e.g. Sokal, 1988) and are currently being investigated. Alternative methodologies are available from quality control analysis (e.g. Hotelling, 1947; Montgomery, 1985; Tang and MacNeill, 1989; Warren, 1990), and these are also being examined. As the application of these procedures to data on forest condition is still at a preliminary stage, emphasis here will be on the identification of anomalous records within 1990, taking no account of data from previous years.

Individual tree data

It would clearly be impossible to go through each of the variables here. However, the data for Sitka spruce, showing the maximum, minimum and mean values, together with standard error and standard deviation, are presented in Table 27. Both the individual tree tables and the site mean tables are used to identify outliers for further examination. In some cases, there is a ready explanation, in others, assistance from the Pathology Diagnostic and Advisory Service is sought.

The process can be illustrated with reference to the crown density scores for Sitka spruce. The mean value obtained was 5.84 and the upper limit (the mean plus twice the standard deviation) was 12.18. This is the class value: its crown density loss equivalent lies between 60% and 65%. Consequently, any tree that had lost 65% of its foliage would be flagged for further examination. A database search indicates that there are 57 such trees, located over 18 sites. Two sites account for 28 trees. By building up lists of problem trees, a shortlist of sites for further investigation can be drawn up, and priority sites identified.

Site mean data

The same process as used for individual tree

Table 27. Basic statistics for individual tree data for Sitka spruce, 1990.

Variable	N	Minimum	Maximum	Mean	Standard error	Standard deviation
Crown density score	1344	0	17.0	5.84	0.09	3.17
Shoot death extent	1344	0	4.0	1.32	0.04	1.35
Secondary shoot abundance	1344	0	4.0	1.14	0.04	1.43
Coning	1344	0	3.0	1.00	0.02	0.91
Current needle browning	1344	0	1.0	0.01	0	0.08
Older needle browning	1344	0	3.0	0.01	0	0.12
Current needle yellowing	1337	0	3.0	0.10	0.01	0.43
Older needle yellowing	1342	0	3.0	0.03	0.01	0.21
Overall discoloration	1344	0	3.0	0.08	0.01	0.35
Mechanical damage %	1344	0	10.0	0.35	0.02	0.88
Butt damage %	1344	0	8.0	0.23	0.02	0.82
Stem damage %	1344	0	6.0	0.04	0.01	0.27
Insect activity	1344	0	4.0	0.95	0.03	1.26
Fungi	1344	0	1.0	0	0	0.03

data can be used here to identify anomalous sites. However, the use of stand mean data has a number of statistical limitations which should be recognised (Schmidtke, 1987). Having pinpointed anomalous sites, they can then be examined and visited by a pathologist or entomologist, if required.

The process can again be illustrated with reference to crown density scores for Sitka spruce. In this case, the mean crown density reduction score was 29.19%, with a standard deviation of 10.57%. This means that any site with a score of 50.33% or greater would be outside the upper limit. Two sites fall above this value, one with a score of 56.2% and the other with a score of 66.7%. Predictably, they are the two sites that accounted for almost half of the individual trees outside the upper limit. Although the use of site means is quicker, it is clearly less sensitive than the individual tree values and it is the latter that development work will concentrate upon.

Environmental factors affecting crown condition

The period from September 1989 to September 1990 was marked by several notable meteorological events. As in 1988–89, the winter was unusually mild, although a cold spell in November was probably responsible for reducing many populations of overwintering insects in the south and west of Britain. Severe winds occurred in January and February, resulting in the loss of many trees in the southern part of the country. New growth in the spring occurred earlier than usual and was badly affected by frosts in early April and late May. In terms of the survey, one of the most notable effects of the frosts was the death of flowers on oak, which resulted in the relatively low occurrence of acorns in the autumn of 1990. Finally, as already mentioned, drought conditions developed over much of England and Wales in the late summer.

The drought in 1990 differed considerably from that of 1989. In 1989, the drought developed early in the summer and extended into August. In 1990, the onset was much later, with many areas that were affected only experiencing unusually dry conditions from the beginning of July onwards.

Ozone levels were relatively high during the summer of 1990, with the number of episode

Table 28. Basic statistics for the stand data for Sitka spruce, 1990.

Variable	N	Minimum	Maximum	Mean	Standard error	Standard deviation
Crown density	56	9.40	66.70	29.19	1.41	10.57
Shoot death extent	56	0	3.21	1.32	0.15	1.11
Coning	56	0	2.33	1.00	0.07	0.53
% trees with needle retention <7 years	56	0	1.00	0.45	0.05	0.35
Current needle browning	56	5.00	7.20	5.09	0.05	0.34
Older needle browning	56	5.00	8.10	5.15	0.07	0.53
Current needle yellowing	56	5.00	33.20	6.95	0.66	4.93
Older needle yellowing	56	5.00	9.70	5.47	0.14	1.08
Overall discoloration	56	5.00	15.70	6.34	0.34	2.52
% trees with crown density >20%	56	0.13	1.00	0.75	0.03	0.22
% trees with crown density >50%	56	0	0.88	0.12	0.02	0.18
% trees with crown density >75%	56	0	0.42	0.02	0.01	0.06
Insect activity	56	0	3.29	0.95	0.14	1.06
Fungi	56	0	0.04	0	0	0.01
Shoot death	56	0	1.00	0.54	0.05	0.36

days (defined as occurring when two or more national network sites experience maximum hourly average ozone concentrations exceeding 60 ppb) being higher than in either 1987 or 1988. In terms of the number of episode days, levels were similar to the summer of 1989, although the timing of the episodes differed (J. S. Bower, personal communication, Warren Spring Laboratory). As with the drought, this difference would have been important for any potential effects on trees. The relationship between crown condition and ozone will be investigated once more detailed information on ozone concentrations in 1990 become available.

Pests and pathogens in 1989–90

Insect pests were not particularly noticeable in the south and west of Britain in 1990. The severe defoliation by *Elatobium abietinum* was not repeated although there were some records of defoliation by this insect at sites unaffected in 1989. Defoliation by *E. abietinum* was apparent at many sites in north Scotland, although sample trees were not particularly affected. The effects were very noticeable on Sitka spruce examined in the European grid survey, with a marked deterioration occurring. The difference between the two suveys may be attributable to the major proportion of Sitka spruce in the grid survey being located in Scotland and to the differences in the ages of trees in the two samples. In the grid survey, the majority of Sitka spruce are very young (< 30 years) whereas in the main survey, the minimum age is 30 years.

The abundance of other insect pests likely to influence the crown condition of the species in the survey was generally unexceptional. *Tortrix viridana* populations were low, although there were some records of defoliation in Cheshire. *Rhynchaenus fagi* populations also appeared to be low over much of England and Wales, and neither 'shot-holes' nor browning was noted as being common.

Discussion

Many of the indices described in this report were only examined for the first time in 1989. Consequently, it has only been possible to compare their development over two years. Others have been available for longer, with the longest records being for the period 1987–90. This is still too short for any conclusions to be drawn on long-term trends in forest condition. A wide range of indices are now recorded and it would be impossible to discuss all of these in full. Consequently, the following discussion is selective and only the most important points have been amplified.

Sitka spruce

The natural range of Sitka spruce is the western seaboard of North America, where it occurs from northern California to Alaska. It only occurs in a narrow coastal belt, never extending more than about 210 km inland (Lines, 1987). Initially seed was derived from a variety of sources, although it seems likely that much came from Washington State. The majority of younger trees are the Queen Charlotte Islands provenance and are therefore strictly maritime in origin. The current distribution of plots (Figure 1a) reflects this trend. Plots are all located within 90 km of the sea and they are generally absent from the south-east of England, which is characterised by a more continental climate.

Over the past 4 years, the most important short-term factor affecting the crown condition of Sitka spruce has been the green spruce aphid. Outbreaks of this pest, brought about by the mild winter of 1988–89 and, to a much lesser extent, the mild winter of 1989–90, caused severe defoliation in Sitka spruce in many areas. Although the impact was noted on many trees involved in the Forestry Commission's main monitoring programme in 1989, it was particularly apparent from the parallel monitoring programme run on behalf of the Commission of the European Communities (CEC).

The defoliation in 1989 was mainly confined to the loss of older needles, although shoot death was recorded in some extreme cases. The defoliation primarily occurred from the base of the crown upwards, or involved the gradual loss of needles from throughout the crown. In 1990, many trees were showing signs of recovery. Unfortunately, it is not possible to directly compare scores of individual trees, as the assessment method in 1989 differed slightly from 1987, 1988 and 1990 (owing to confusion over the precise definition of the 5% classes). However, the overall figures (Table 8c) indicate that a significant improvement occurred (Kolmogorov-Smirnov two-tailed test, $p < 0.05$).

The crown densities of individual Sitka spruce have fluctuated considerably between 1987 and 1989, although there has been a general tendency for lower (i.e. better) scores. The cause for this general improvement is unknown, but it seems likely that the absence of any major climatic stresses in the area in which Sitka spruce occurs may have played a major part. The reason for the high defoliation scores at the start of the programme is also unknown, and is being investigated through dendroecological methods.

Dead shoots were recorded on over half of the trees, being common or abundant on 30%. The cause of death was not recorded, but a variety of factors could be involved.

As in previous years, discoloration was relatively rare. Browning was present on 1% of the trees, although yellowing was slightly more frequent, with 5% of trees having yellowing of current-year needles. The amount of browning represents a reduction on 1989, suggesting that the factor causing the browning in 1989 was ineffective or not present in 1990. Yellowing of older needles was also less frequent, although it is unknown whether these needles were dropped or recovered. The increase in the amount of yellowing of current-year needles is of interest. There was no apparent pattern to the increase, with the sites showing the greatest increases being located in Cornwall, Anglesey and Speyside. A variety of nutrient disorders can cause yellowing and, without foliar analysis, it is impossible to assign a cause.

The values for needle retention apparently contradict the crown density results, with many trees showing a reduction in needle retention. The reductions were most apparent in trees with high retention values in 1989. As needle retention is related to crown density, it can be inferred that the reductions occurred primarily in well-foliaged trees, whereas the improvements occurred in severely defoliated trees.

A marked increase in the amount of coning was seen in Sitka spruce in 1990. Only 36% of trees were without coning, compared to 69% in 1989. There was no indication that the amount of coning was related to crown density, although severely defoliated trees were generally without cones.

The results for Sitka spruce are generally encouraging, although the substantial fluctuations that occur from year-to-year clearly warrant careful monitoring. In the short span of data that is available, there is no indication of any downward trend in condition, although such a trend would be difficult to identify in view of the effects of the spruce aphid. Examination of anomalous plots has not revealed any unexplained problems, although the precise cause-effect mechanisms have not been evaluated for the two plots in worst condition.

Norway spruce

Norway spruce is the species of greatest interest in central Europe in relation to the problem of forest decline. A number of different types of decline have been recognised on the basis of their distributions and symptoms (Forschungsbeirat Waldschaden, 1986). The most important form is known as Type 1 decline and is associated with yellowing of older needles, caused by magnesium deficiency, and needle loss (Roberts *et al.*, 1989). Innes and Boswell (1990) have argued that this type of defoliation, and most of the other types recorded in Germany, has not been detected in Britain, with defoliation taking a very different form. Consequently, different cause-effect complexes appear to be present in Britain.

Throughout the past 4 years, the crown density of Norway spruce has remained remarkably constant and individual trees appear to have been more stable than Sitka spruce. Norway spruce is not subject to *Elatobium* defoliation episodes to the same extent as Sitka spruce and, as it tends to be planted in more sheltered situations, it appears to be less susceptible to severe climatic stress. It is not as sensitive to drought as some species and a complex set of conditions appear to be necessary for drought to trigger any decline in condition (Cramer and Cramer-Middendorf, 1984; Munster-Swendsen, 1987). The number of common sample trees for the period 1987–90 is the highest of any of the species and the figures are therefore likely to be the most reliable. However, it is worth noting that the storms of 1987 and 1990 resulted in the loss of the majority of Norway spruce sites in the south-east of England, and this area continues to be under-represented.

The majority of defoliation takes the form of a uniform loss of needles throughout the crown, with loss of needles from the base of the crown upwards being the second most frequent form. The unexplained syndrome known as 'top-dying' was particularly noticeable in 1990, although very few of the sample plots were affected. As 'top-dying' does not necessarily involve dying from the top downwards (browning of foliage is often the more obvious symptom), the presence of the problem in sample plots is probably better judged from symptoms other than defoliation type.

Shoot death was marginally less frequent in Norway spruce than in Sitka spruce but, as there were problems with the assessment of this index, any comparisons should be made with care. Approximately half of the sample trees were recorded as having shoot death present, and shoot death was common in 23%. Shoot death is likely to occur naturally through suppression (although shoot death caused by suppression should have been excluded from the assessment) and there may also be some turnover of shoots within the crown. Little is known about the phenomenon and the monitoring programme is likely to reveal the nature of any turnover of shoots within the crown and how this affects other parameters of crown condition.

As already indicated, needle discoloration is an important symptom associated with most of the different types of Norway spruce decline that have been identified. Very little discoloration was noted in Norway spruce although the repeat assessments suggest that trees located on drought-prone sites might have shown more discoloration towards the end of the survey period. The small number of records of yellowing may well be related to the preponderance of sites being located away from the areas affected by the 1990 drought. A small increase in the amount of browning of current needles was noted, but browning of older needles was generally less frequent than in 1989. Yellowing of older needles was also less frequent.

Needle retention was lower in 1990 than in 1989. Many trees scored in 1989 as having more than 7 years' needle retention had 7 years or less in 1990. An explanation for this, particularly in view of the absence of any marked effects on crown density, is not available.

Leaders were absent on a relatively high proportion (16%) of trees, the proportion having increased since 1989. There was little evidence of side shoots having taken over; 21 trees recorded as having missing leaders in 1989 were recorded as having a side shoot that had taken over as a leader in 1990.

The climatic conditions over the past 24 months are undoubtedly responsible for the increase in the amount of coning recorded in 1990. 46% of the trees recorded in 1989 as being without coning had cones in 1990 and, in 3% of the cases, coning was abundant. The amount of coning was less than in Sitka spruce, but the extent is still of note. The amount of coning may well influence the condition and growth of trees in 1991 and 1992 and this will be carefully monitored.

The general picture for Norway spruce is one of a fairly stable situation, with only a few indices fluctuating to any degree. There is no indication of any major nutrient problems in the sample trees, which would be apparent from foliage discoloration.

Scots pine

Scots pine is the most important native conifer in Britain, although some of the provenances used are exotic. Problems with pine have been identified in several areas, notably the

Netherlands and northern Germany, where excessive deposition of ammonia is believed to be causing problems (den Boer and van den Tweel, 1985; van Breemen and van Dijk, 1988; van Dijk and Roelofs, 1988; Roelofs et al., 1987; de Temmerman and Coosemans, 1989). Growth disturbances, foliage discoloration and increased insect activity have all been reported from affected trees. Ammonia deposition is also high in some parts of Britain, although precise figures remain elusive because of difficulties in the measurement of some forms of deposition (Williams et al., 1989).

Of the five species, examined, the data for Scots pine showed the least consistency between observers. Care should therefore be taken in interpreting the following.

There was little change in the crown density of Scots pine between 1989 and 1990 although, over the past 4 years, considerable changes have occurred. Scots pine remains the species with highest recorded mortality in the survey, with 13 trees (0.7%) being recorded as dead in 1990. Six of these are at a single site, which has previously been identified as having a major problem (Innes and Boswell, 1989). However, a satisfactory explanation for the decline and death of the trees at this and other similar sites has yet to be found. There was an increase in mortality in 1990; this took the form of single trees at a number of sites. The factors leading to the mortality are unknown.

The crown density data for Scots pine illustrate the complex pattern of improvement and deterioration that makes up the annual figures for forest condition. For example, although an increase in mortality was identified in 1990, some of the trees with moderate to severe defoliation improved. Over the past 4 years, there has been a general improvement in crown density. In some cases, this has been very marked, with trees scored as having between 71% and 80% loss of density in 1987 being scored as having less than 10% loss of density in 1990. Relatively rapid changes would be expected in Scots pine as a single year of needles makes up 33% to 50% of the foliage in the crown and, over a 4-year period, a complete turnover of the foliage would be expected to occur.

As with the spruce species, the most frequent forms of defoliation involve uniform thinning throughout the crown and thinning from the base upwards. Shoot death was recorded marginally more frequently in Scots pine than in the two spruce species, but the numbers of dead shoots on individual trees were generally less.

Needle discoloration was also more frequent on Scots pine. The changes from 1989 are difficult to interpret as new discoloration was matched by almost equal numbers of trees showing recovery. The greatest overall change was in the amount of yellowing of older needles, which showed a marked reduction. The high numbers of trees developing browning on the current needles suggests that trees were under a similar stress to that which caused the browning of younger needles in 1989. Drought might seem the obvious candidate, but only 23 trees were recorded as having browning of current needles in both years. This compares to 126 trees which developed the symptom and 102 trees which recovered. As the droughts in 1989 and 1990 occurred in similar parts of Britain, they are unlikely to be the cause of the browning. The most likely explanation is the death of shoots caused by the pine shoot beetle (*Tomicus piniperda* L.; Coleoptera) since most of the records of browning were for the majority of needles on a few individual shoots rather than a few needles on each shoot.

Overall, needle retention in 1990 was very similar to retention in 1989. However, as with some of the other indices, this masks considerable changes in individual trees. For example, 236 trees with 2 years of needles in 1989 had 3 years in 1990 whereas 213 trees with 3 years in 1989 had 2 years in 1990. The proportion of trees with only one year of needles remained very low (1%).

The amount of flowering in the crown also remained similar overall, although there was a slight reduction in the amount of flowering in the upper crown. Considerable changes occurred on individual trees, making any pattern identification difficult. As in 1989, heavy flowering was associated with lower crown densities. Coning was rather less frequent in 1990, in marked contrast to the two spruce species.

Scots pine showed much more mechanical damage to the crowns than either of the two spruce species. Damage mainly took the form of abrasion to peripheral shoots and, as with the spruces, there was a quite marked increase in the amount of abrasion as compared to 1989. This is hardly surprising given the storms of early 1990. Structural damage to the crowns caused by wind was also more apparent in Scots pine, with 9% of trees affected.

Overall, Scots pine showed little change from 1989. The amount of mortality remains higher than in the other species, but half of this is accounted for by one site. The crown densities are gradually improving on 1987 levels, although wind and insect damage can cause local problems. There is little evidence that the trees have been adversely affected by the two droughts, confirming the general impression that Scots pine is a drought-resistant species.

Oak

Oak is the commonest broadleaved species in Britain and, as such, it is of considerable commercial, ecological and aesthetic value. There has been some concern about the state of hedgerow oaks in Britain, but these are excluded from this monitoring programme. There has also been concern about an apparent increase in the amount of dieback in forest oaks in Europe (Hartmann et al., 1989). In 1989 and 1990, there were reports of similar dieback in Britain, and these are currently the subject of a special study by the Forestry Commission's Pathology Branch.

The sample size for oak has been gradually increasing over the past 4 years, and the current figure of 73 sites is approaching the target of 80. 1990 saw a substantial increase in the number of sites (22 new plots), and more extensive data are therefore available for 1990 compared to 1989.

Of all species investigated, oaks have the poorest scores. This is not considered to indicate that oaks are more damaged; the amount of ground vegetation under oak is well-known and reflects the amount of light that is normally able to penetrate the canopy. The crown density of oak has been rather variable since 1987, although currently there are more trees in the lowest three defoliation classes than in any other year. The number of severely defoliated trees is also at its lowest level since observations began. A relatively high proportion of trees were in the same defoliation category in 1990 as they were in 1987, although the improvement in condition of the poorer trees is also very noticeable.

The majority of defoliation took the form of small gaps in the crown or small gaps in the lateral branch system, although a significant number of trees had large gaps present. The figures for 1989 and 1990 were very similar.

Dieback provides a relatively good indication of the condition of a broadleaved tree. Only 33% of oaks were without dieback, with small amounts of dieback being recorded in 43% of the trees. In many cases, the dieback was noted throughout the crown. The majority of cases involved a relatively small proportion of the crown: 96% of the trees had less than 20% of the crown affected. The number of trees affected by dieback increased during the year, a trend that may be related to the droughts.

Discoloration of foliage was relatively rare, involving less than 10% of the trees. Yellowing was more frequent than browning, and there were a number of cases of moderate to severe discoloration. As with the conifers, the trees affected were different in 1989 and 1990.

In 1989, a feature of oak was the very heavy acorn crop. This was not repeated in 1990, and the level of fruiting was lower than in any of the other species. The reduction in the numbers of acorns is believed to be due to the late spring frosts that occurred over much of Britain in 1990. In many cases, the frosts killed the flowers of oak, resulting in the failure of trees to set acorns.

Levels of mechanical damage to the crowns of oak were very similar in 1989 and 1990, although the proportion of trees recorded as having wind damage increased. Abrasion of peripheral shoots was more apparent on oak than on any of the other species.

Although the crown density of many oaks improved, the increase in minor dieback suggests that many trees were under stress. The cause of the increase in dieback is unclear: both drought and frost damage may have been involved. The development of severe dieback has been noted in some British oaks (including some of the sample trees) and this warrants careful investigation.

Beech
There has been considerable concern over the condition of beech in Europe and this concern has extended to Britain. Much of the interest has arisen as a result of the work on shoot growth by Andreas Roloff (Roloff, 1985a, 1985b; Roloff and Linnard, 1985). Studies in Britain (Lonsdale, 1986a, 1986b; Lonsdale et al., 1989) suggest that the shoot growth depressions reported in Germany are also present in Britain, although the causes are by no means certain. These studies indicate that twig extension growth at some sites was severely depressed by the 1976 drought and that, in some cases, growth has not recovered to pre-1976 levels.

The monitoring programme currently involves the assessment of 36 beech plots, although this figure is likely to be more than doubled in 1991, bringing the total to approximately 80 plots. The small sample size, the marked changes in tree condition that occurred during the course of the survey and the problems over the consistency of some of the data all combine to make any interpretation of the results difficult. An increased sample size and further improvements in observer consistency should help interpretation in the future.

In contrast to the other species, which either improved or remained constant, the crown densities of beech generally deteriorated between 1989 and 1990. However, the proportion of trees in the lowest defoliation class was still well above the proportion in either 1987 or 1988. Conversely, considerably more severely defoliated trees were present. In most cases, the deterioration involved individual trees that had previously been scored as having relatively low levels of defoliation. Reductions on 1989 values of up to 70% were recorded, indicating that some trees were severely stressed.

As with the oak, the majority of defoliated trees were classed as having either small gaps in the foliage or gaps in the lateral branch systems. However, the proportions of trees in these two categories showed little change from 1989. 48% of the trees were scored in Roloff's category 1 and a further 21% in category 2.

Crown dieback was more frequent in beech than in oak, having increased markedly in 1990. It primarily involved leaf loss or death of small branches and was mostly restricted to the top or the top and middle of the crowns. As with oak, it mainly involved a relatively small (< 20%) proportion of the crown. There was evidence of a progression of dieback, with almost half of the trees scored in 1989 as having leaf loss only being scored in 1990 as having dieback involving small branches. However, a considerable number of trees without dieback in 1989 moved to the small branch category in 1990. Many of the trees with some degree of dieback in 1989 showed a reduction in the proportion of the crown affected in 1990, indicating that the changes were far from uniform. Further dieback may be expected in 1991 and 1992 following the heavy masting of 1990 (Skelly et al., 1987).

There was a marked reduction in the amount of both browning and yellowing of leaves. There was less browning than in any of the previous 3 years and less yellowing than in either 1988 or 1989. These changes contrast with the other indices, emphasising the need for recording a variety of measures in order to characterise crown condition adequately.

There was a major increase in the amount of mast in 1990. Very often, this was associated with unusually small leaves, which were also much more frequent in 1990. Only 17 trees showed a reduction in the amount of mast on 1989 levels, whereas 707 (86%) showed an increase. The degree of masting was so high in some cases that it may have affected the crown density estimates. Masting was particularly high on moderately to severely defoliated trees, as has been recorded in Germany (Gartner, 1988), suggesting that the increase was in response to stress. Heavy fruiting is known to occur in response to previous summer growing season drought (Bernier *et al.*, 1989), and this seems to be the most likely stress involved in the triggering of the high mast load.

Leaf-rolling also increased in 1990, although not by the same extent as masting. Leaf-rolling is likely to reflect the degree of stress experienced by trees during the current growing season, whereas the amount of mast is determined during the previous growing season. The severity of leaf-rolling was very similar in 1989 and 1990.

The amount of mechanical damage in beech crowns increased, with wind damage being the main factor involved. Both butt and stem damage were less evident than in 1989.

A number of indices of crown condition in beech changed between 1989 and 1990. The changes were relatively consistent and strongly suggest that climatic conditions were important. The effects of the 1989 were reflected by increased levels of dieback, small leaves and masting, whereas the effects of the 1990 drought were evident in the amount of leaf-rolling and premature leaf loss. Structural damage has occurred in some trees, presumably as a result of the 1990 storms. The reduction in leaf discoloration appears to be anomalous. However, browning of beech leaves in Britain is primarily caused by the beech leaf miner (*Rhynchaenus fagi* L.; Coleoptera) and the fungus *Apiognomonia errabunda* (Roberge) Hohnel, both of which appeared to be less frequent in 1990. Their low incidence may be associated with changes in the mineral composition of the leaves (particularly the nitrogen content) which are known to occur during years of heavy seed production (Bernier and Brazeau, 1988; Bernier *et al.*, 1989). Yellowing develops on particular sites in most years, and the reasons for the relatively low levels in 1990 are unknown. The changes in the condition of individual trees were very variable, reflecting the known variation in the response of beech to environmental stress (Muller-Starck, 1985).

Conclusions

As in previous years, the condition of trees in Britain showed considerable variation. The 1990 results support the contention that tree condition fluctuates around a mean, with annual variations depending on extraneous factors such as climate and insect activity (which themselves are linked). Others have also reached this conclusion (e.g. Kandler, 1988, 1989). There is no indication of any long-term deterioration in tree condition, although it is highly unlikely that such a trend would become evident over the space of 4 years. Instead, the short time series that is available indicates that tree condition is generally improving, a trend that is supported by data from elsewhere in Europe (Mettendorf *et al.*, 1988; Anon., 1989b; Cerny, 1989).

The successive droughts of 1989 and 1990 have clearly had an impact on some trees, although this is not reflected in the overall figures for four of the five species. The drought in 1990 occurred during the course of the monitoring programme, and repeat assessments indicated that the condition of individual trees changed between early July and late August. Consequently, the full effects of the 1990 drought are unlikely to have been recorded. While this presents a limitation for evaluating the effects of the 1990 drought, in the longer term, it means that the amount of 'noise' associated with the 1990 drought will be lessened. As in 1989 (Innes *et al.*, 1989), the effects of the drought were more apparent in isolated trees than forest trees. The reasons for this have not been formally established although it seems likely that the microclimatic conditions in forests during drought periods are more favourable than around isolated trees.

The storms of January and February 1990 caused considerable damage to trees in southern Britain, but the damage to plots included in the monitoring programme was much less obvious than in 1987, when the storm occurred while the trees were still in leaf. A few sites were lost. Structural damage and peripheral damage to shoots was recorded. The results from 1988 and 1989 suggest that the trees will gradually recover from this damage.

The severe defoliation of Sitka spruce that occurred as a result of infestations by the green spruce aphid was mostly absent although there were some records of defoliation at sites unaffected in 1989. Most of the trees badly affected by the outbreak were showing signs of recovery at the time of the assessments and, in the absence of any repeat outbreaks, further improvements can be expected.

The problems created by the changing condition of trees during the survey programme and the increased level of variation in the scoring of several of the indices meant that some of the analyses undertaken in previous years were inappropriate in 1990. Consequently, no maps of specific indices have been drawn up, nor has any attempt been made to relate indices of crown condition to environmental factors on a site by site basis. This is seen as a temporary problem which it should be possible to resolve in 1991, when a larger sample size will also be available.

The analysis of the data is becoming increasingly sophisticated as improved computing facilities become available and as data are added to the database. It is now possible to identify both anomalous trees and anomalous sites quickly and with a high degree of confidence, making rapid follow-up visits possible. The progress of individual trees can be monitored and related to specific events affecting that tree. The environmental database is steadily being improved and information on pollutant deposition for the period 1986 to 1988 and monthly climatic conditions over the period April 1984 to the present have recently been added. A new method of analysis, based on multivariate techniques has shown considerable promise and will be used to analyse the 1991 data, provided that they are more reliable than the 1990 data. In addition, space-time techniques appear to have considerable potential and are currently being examined.

Over the past 4 years the Forestry Commission has developed considerable expertise in the assessment and analysis of data on forest condition and the forest plots currently represent one of the most extensive environmental monitoring networks in Europe. While gaseous air pollution and acidic deposition appear to be having relatively little impact on forests in Britain, the monitoring network represents a major source of information on the effects of environmental change. Consequently, the monitoring is not only being continued, but the density of sampling is being increased. It is planned that the network will be augmented in the next 12 months by a small number of detailed monitoring sites, where intensive monitoring will be conducted. The two networks will provide the baseline data without which a balanced and accurate picture of trends in forest condition would be impossible.

ACKNOWLEDGEMENTS

The field assessments were made by staff from Forest Surveys Branch, without whose assistance the work would have been impossible. Their continuing dedication and commitment to the programme is deeply appreciated. Lesley Halsall and John Hall together produced the data collection programme for the hand-held microcomputers at extremely short notice, greatly improving the data capture and transfer rate. Members of Mensuration Branch provided much-needed advice and support during the course of the monitoring programme and subsequent analysis.

REFERENCES

ANON. (1989a). *Patterns of forest condition and air pollution: a report of the progress of the National Vegetation Survey*. United States Department of Agriculture, Forest Service, Southeastern Forest Experiment Station.

ANON. (1989b). *Waldzustandsbericht. Ergebnisse der Waldschädenserhebung 1989*. Bundesministerium für Ernahrung, Landwirtschaft und Forsten, Bonn.

ATHARI, S. and KRAMER, H. (1989). Problematik der Zuwachsuntersuchungen in Buchenbestanden mit neuartigen Schadsymptomen. *Allgemeine Forst- und Jagdzeitung* **160**, 1-8.

BALDER, H. and LAKENBERG, E. (1987). Neuartiges Eichensterben in Berlin. *Allgemeine Forst Zeitschrift* **42**, 684-685.

BECHER, G. (1986). Ergebnisse und methodisch-theoretische Uberlegungen zur immissions-okologischen Waldzustandserfassung (IWE) – dargestellt am Beispiel Hamburgs –. *Forstarchiv* **57**, 167-174.

BERNIER, B. and BRAZEAU, M. (1988). Foliar nutrient status in relation to sugar maple dieback and decline in the Quebec Appalachians. *Canadian Journal of Forest Research* **18**, 754-761.

BERNIER, B., PARE, D. and BRAZEAU, M. (1989). Natural stresses, nutrient imbalances and forest decline in southeastern Quebec. *Water, Air, and Soil Pollution* **48**, 239-250.

den BOER, W.M.J. and van den TWEEL, P.A. (1985). The health condition of the Dutch forests in 1984. *Netherlands Journal of Agricultural Science* **33**, 167-174.

BOSSHARD, W. (1986). *Kronenbilder*. Eidgenössische Anstalt für das forstliche Versuchswesen, Birmensdorf.

van BREEMEN, N. and van DIJK, H.F.G. (1988). Ecosystem effects of atmospheric deposition of nitrogen in the Netherlands. *Environmental Pollution* **54**, 249-274.

ČERNÝ, M. (1989). Současný zdravotní stav jedle bělokoré na území ČSSR. *Lesnická Práce* **68**, 402-407.

CLINE, S.P. and BURKMAN, W.G. (1989). The role of quality assurance in ecological research programs. In *Air pollution and forest decline*, eds J.B. Bucher and I. Bucher-Wallin, 361-365. Eidgenössische Anstalt für das forstliche Versuchswesen, Birmensdorf.

CRAMER, H.H. and CRAMER-MIDDENDORF, M. (1984). Studies on the relationships between periods of damage and factors of climate in the Middle European forests since 1851. *Pflanzenschutz Nachrichten Bayer* **37**, 208-334.

van DIJK, H.F.G. and ROELOFS, J.G.M. (1988). Effects of excessive ammonium deposition on the nutritional status and condition of pine needles. *Physiologia Plantarum* **73**, 494-501.

DOBLER, D., HOHLOCH, K., LISBACH, B. and SALIARI, M. (1988). Trieblangen-Messungen an Buchen. *Allgemeine Forst Zeitschrift* **43**, 811-812.

DONAUBAUER, E. (1987). Auftreten von Krankheiten und Schadlingen der Eiche und ihr Bezug zum Eichensterben. *Österreichische Forstzeitung* **98**, 46-48.

EICHHOLZ, U. (1985). Sterben von Eichenjungbestanden in Südhessen. *Allgemeine Forst Zeitschrift* **40**, 47-48.

FORSCHUNGSBEIRAT WALDSCHÄDEN (1986). 2. Bericht. Karlsruhe.

GARTNER, E.J. (1988). Operational forest decline symptomatology. In *Scientific basis of forest decline symptomatology*, eds J.N. Cape and P. Mathy, 295-306. Commission of the European Communities, Air Pollution Report 15, Brussels.

GRUBER, F. (1987). *Der Verzweigungssystem und der Nadelfall der Fichte (Picea abies (L.) Karst.) als Grundlage zur Beurteilung von Waldschäden*. Beiträge Forschungs- zentrum Waldokosystemen/Waldschäden A26. (214 pp.)

HAGNER, S. (1956). Om kott- och froproduktionen i svenska barrskogar. *Meddelanden fran Statens Skogsforskningsinstitut* **47**, No. 8, 1-117.

HARTMANN, G., BLANK, R. and LEWARK, S. (1989). Eichensterben in Norddeutschland – Verbreitung, Schadbilder, mogliche Ursachen. *Der Forst und Holz* **44**, 475-487.

HOTELLING, H. (1947). Multivariate quality control. In *Techniques of statistical analysis*, eds Eisenhart, Hastay and Wallis. John Wiley, New York.

INNES, J.L. (1990). *Assessment of tree condition*. Forestry Commission Field Book 12. HMSO, London.

INNES, J.L. and BOSWELL, R.C. (1987). *Forest health surveys 1987. Part 1: results*. Forestry Commission Bulletin 74. HMSO, London.

INNES, J.L. and BOSWELL, R.C. (1989). *Monitoring of forest condition in the United Kingdom 1988*. Forestry Commission Bulletin 88. HMSO, London.

INNES, J.L. and BOSWELL, R.C. (1990a). *Monitoring of forest condition in Great Britain 1989*. Forestry Commission Bulletin 94. HMSO, London.

INNES, J.L. and BOSWELL, R.C. (1990b). Reliability, presentation, and relationships among the data from inventories of forest condition. *Canadian Journal of Forest Research* **20**, 790-799.

INNES, J.L., BOSWELL, R.C. and LONSDALE, D. (1989). *Weather conditions during the summer of 1989 and their effect on trees.* Research Information Note 162. Forestry Commission, Edinburgh.

JAKUCS, P. (1988). Ecological approach to forest decay in Hungary. *Ambio* **17**, 267-274.

JUKOLA-SULONEN, E.-L., MIKKOLA, K., NEVALAINEN, S. and YLI-KOJOLA, H. (1987). Havupuiden elinvoimaisuus Suomessa 1985-6. *Metsantutkimuslaitoksen Tiedonantoja* **256**, 1-92.

KANDLER, O. (1988). Epidemiologische Bewertung der Waldschädenserhebungen 1983 bis 1987 in der Bundesrepublik Deutschland. *Allgemeine Forst- und Jagdzeitung* **159**, 179-194.

KANDLER, O. (1989). Epidemiological evaluation of the course of 'Waldsterben' from 1983 to 1987. In *Air pollution and forest decline*, eds J.B. Bucher and I. Bucher-Wallin, 297-302. Eidgenössische Anstalt für das forstliche Versuchswesen, Birmensdorf.

LESINSKI, J.A. (1989). Dynamics of injury symptoms in Norway spruce. In *International congress on forest decline research: state of knowledge and perspectives*. Poster Abstracts Vol. 1, 55-56. Forschungsbeirat Waldschäden/Luftverunreinigungen der Bundesregierung und der Lander, Bonn.

LESINSKI, J.A. and LANDMANN, G. (1988). Crown and branch malformation in conifers related to forest decline. In *Scientific basis for forest decline symptomatology*, eds J.N. Cape and P. Mathy, 92-106. Commission of the European Communities, Air Pollution Report 15, Brussels.

LESINSKI, J.A. and WESTMAN, L. (1987). Crown injury types in Norway spruce and their applicability for forest inventory. In *Acid rain: scientific and technical advances*, eds R. Perry, R.M. Harrison, J.N.B. Bell and J.N. Lester, 657-662. Selper, London.

LINES, R. (1987). *Choice of seed origins for the main forest species in Britain.* Forestry Commission Bulletin 66. HMSO, London.

LONSDALE, D. (1986a). *Beech health study 1985.* Forestry Commission Research and Development Paper 146. Forestry Commission, Edinburgh.

LONSDALE, D. (1986b). *Beech health study 1986.* Forestry Research and Development Paper 149. Forestry Commission, Edinburgh.

LONSDALE, D., HICKMAN, I.T., MOBBS, I.D. and MATTHEWS, R.W. (1989). A quantitative analysis of beech health and pollution across southern Britain. *Naturwissenschaften* **76**, 571-573.

MAHRER, F. (1989). Problems in the determination and interpretation of needle and leaf loss. In *Air pollution and forest decline*, eds J.B. Bucher and I. Bucher-Wallin, 229-231. Eidgenössische Anstalt für das forstliche Versuchswesen, Birmensdorf.

METTENDORF, B., SCHROTER, H. and HRADETSKY, J. (1988). Analysenergebnisse zur Schadensentwicklung auf Tannen- und Fichten-Dauerbeobachtungsflächen in Baden-Württemberg. *Allgemeine Forst- und Jagdzeitung* **159**, 171-177.

MONTGOMERY, D.C. (1985). *Statistical quality control.* McGraw-Hill, New York.

MULLER-STARCK, G. (1985). Genetic differences between 'tolerant' and 'sensitive' beeches (*Fagus sylvatica* L.) in an environmentally stressed adult forest stand. *Silva Genetica* **34**, 241-247.

MÜNSTER-SWENDSEN, M. (1987). Sammenhaeng imellem nedbør, tilvaekst og viklerangreb hos rødgran. *Dansk Skovforenings Tidsskrift* **72**, 41-50.

RAUBERTAS, R. (1988). Spatial and temporal analysis of disease occurrence for detection of clustering. *Biometrics* **44**, 1121-1129.

RAUBERTAS, R.F., BROWN, P., CATHALA, F. and BROWN, I. (1989). The question of clustering of Creutzfeldt-Jakob disease. *American Journal of Epidemiology* **129**, 146-154.

REHFUESS, C. and REHFUESS, K.E. (1988). Ersatztriebe an Fichten — Entwicklung, Nahrstoffversorgung und Bedeutung für die Kronenmorphologie. *Allgemeine Forst- und Jagdzeitung* **159**, 20-26.

REHFUESS, K.E. (1989). Acidic deposition — extent and impact on forest soils, nutrition, growth and disease phenomena in central Europe: a review. *Water, Air, and Soil Pollution* **48**, 1-20.

ROBERTS, T.M., SKEFFINGTON, R.A. and BLANK, L.W. (1989). Causes of Type 1 spruce decline in Europe. *Forestry* **62**, 179-222.

ROELOFS, J.G.M., BOXMAN, A.W., van DIJK, H.F.G. (1987). Effects of airborne ammonium on natural vegetation and forests. *Ammonia and acidification, Proceedings of a EURASAP Symposium*, 266-276.

ROLOFF, A. (1985a). Schadstufen bei der Buche. *Der Forst und Holzwirt* **40**, 131-134.

ROLOFF, A. (1985b). Untersuchungen zum vorzeitigen Laubfall und zur Diagnose von Trockenschaden in Buchenbeständen. *Allgemeine Forst Zeitschrift* **40**, 157-160.

ROLOFF, A. and LINNARD, W. (1985). Auswirkung von Immissionsschaden in Buchenbeständen. *Allgemeine Forst Zeitschrift* **40**, 905-908.

SCHLAEPFER, R. (1985). Problems in the planning of forest damage inventories. In *Inventorying and monitoring endangered forests*, ed. P. Schmid-Haas, 339-342. Eidgenössisische Anstalt für das forstliche Versuchswesen, Birmensdorf.

SCHLAEPFER, R., MANDALLAZ, D., COMMARMOT, B., GUNTER, R. and SCHMID, B. (1985). Der Gesundheitzustand des Waldes im Revier Schaffhausen. Zur Methodik und Problematik der Erhebung auf Betriebsebene. *Schweizerische Zeitschrift für Forstwesen* **136**, 1-18.

SCHMIDTKE, H. (1987). Zur Definition von Schadniveaus für Waldschädensinventuren. *Allgemeine Forst Zeitschrift* **42**, 930-932.

SHAW, G.M., SELVIN, S., SWAN, S.H., MERILL, D.W. and SCHULMAN, J. (1988). An examination of three spatial disease clustering methodologies. *International Journal of Epidemiology* **17**, 913-919.

SKELLY, J.M., DAVIS. D.D., MERRILL, W. CAMERON, E.A., BROWN, H.D., DRUMMOND, D.B. and DOCHINGER, L.S. (1987). *Diagnosing injury to eastern forest trees*. United States Department of Agriculture, Forest Service, Forest Pest Management, Atlanta, Georgia and the Pennsylvania State University, University Park, Pennsylvania.

TANG, S.M. and MACNEILL, I.B. (1989). The effect of autocorrelated errors on change-detection statistics. *Environmental Monitoring and Assessment* **12**, 203-226.

de TEMMERMAN, L. and COOSEMANS, P. (1989). Ammonia and ammonium deposition on pine stands in Belgium. In *Air pollution and forest decline*, eds J.B. Bucher and I. Bucher-Wallin, 91-96. Eidgenössische Anstalt für das forstliche Versuchswesen, Birmensdorf.

THIEBAUT, B. (1988). Tree growth, morphology and architecture, the case of beech: *Fagus sylvatica* L. In *Scientific basis of forest decline symptomatology*, eds J.N. Cape and P. Mathy, 49-72. Commission of the European Communities, Air Pollution Report 15, Brussels.

WARREN, W.G. (1990). Some novel statistical analyses relevant to the reported growth decline of pine species in the Southeast. *Forest Science* **36**, 448-463.

WESTMAN, L. (1989). A new method for assessing of visible damage to birch and other deciduous trees. In *Air pollution and forest decline*, eds J.B. Bucher and I. Bucher-Wallin, 223-228. Eidgenössische Anstalt für das forstliche Versuchswesen, Birmensdorf.

WILLIAMS, M.L., ATKINS, D.H.F., BOWER, J.S., CAMPBELL, G.W., IRWIN, J.G. and SIMPSON, D. (1989). *A preliminary assessment of the air pollution climate of the UK*. Report LR 723 (AP), Warren Spring Laboratory, Stevenage.